Women in the History of Philosophy and Sciences

Volume 25

As the historical records prove, women have long been creating original contributions to philosophy. We have valuable writings from female philosophers from Antiquity and the Middle Ages, and a continuous tradition from the Renaissance to today. The history of women philosophers thus stretches back as far as the history of philosophy itself. The presence as well as the absence of women philosophers throughout the course of history parallels the history of philosophy as a whole.

Edith Stein, Hannah Arendt and Simone de Beauvoir, the most famous representatives of this tradition in the twentieth century, did not appear from nowhere. They stand, so to speak, on the shoulders of the female titans who came before them.

The series Women in the History of Philosophy and Science published by Springer is of interest not only to the international philosophy community, but also for scholars in history of science and mathematics, the history of ideas, and in women's studies.

Mette Lebech
Editor

Stein and Husserl's Phenomenology of the Social in *Ideas II*

Springer

Editor
Mette Lebech
Maynooth University
Maynooth, Ireland

ISSN 2523-8760 ISSN 2523-8779 (electronic)
Women in the History of Philosophy and Sciences
ISBN 978-3-032-17338-6 ISBN 978-3-032-17339-3 (eBook)
https://doi.org/10.1007/978-3-032-17339-3

This Springer imprint is published by the registered company Springer Nature Switzerland AG
The registered company address is: Gewerbestrasse 11, 6330 Cham, Switzerland

To Jacob Dahl Rendtorff

Series Foreword

Women in the History of Philosophy and Sciences

The history of women's contributions to philosophy and the sciences dates back to the very beginnings of these disciplines. Theano, Hypatia, Du Châtelet, Agnesi, Germain, Lovelace, Stebbing, Curie, Stein are only a small selection of prominent women philosophers and scientists throughout history.

The Springer Series *Women in the History of Philosophy and Sciences* provides a platform for publishing cutting-edge scholarship on women's contributions to the sciences, to philosophy, and to interdisciplinary academic areas. We therefore include in our scope women's contributions to biology, physics, chemistry, and related sciences. The Series also encompasses the entire discipline of the history of philosophy since antiquity (including metaphysics, aesthetics, philosophy of religion, etc.). We also welcome work about women's contributions to mathematics and to interdisciplinary areas such as philosophy of biology, philosophy of medicine, sociology, etc.

The research presented in this series serves to recover women's contributions and to revise our knowledge of the development of philosophical and scientific disciplines, so as to present the full scope of their theoretical and methodological traditions. Supported by an advisory board of internationally esteemed scholars, the volumes offer a comprehensive, up-to-date source of reference for this field of growing relevance. See the listing of planned volumes.

The Springer Series *Women in the History of Philosophy and Sciences* will publish monographs, handbooks, collections, anthologies, and dissertations.

Editors-in-Chief

Paderborn, Germany — Ruth Edith Hagengruber
Toronto, Ontario, Canada — Antonio Calcagno
La Plata, Argentina — Silvia Manzo

Acknowledgements

Ruth Hagengruber and the other Series Editors graciously proposed this collection of articles to appear in their important series, bringing out things 'forgotten' with a gender dimension. I am very grateful for this. Ruth Hagengruber also, as Director of the Centre for the History of Women Philosophers and Scientists at the University of Paderborn, hosted the conference from which many of the contributions of this volume stem.

I am also grateful to Maynooth University for the sabbatical leave that allowed me to initiate the project and take some of the steps towards its realisation. Maynooth University likewise funded a hybrid follow-up seminar through its EDI office: 'Taking Stein Studies to the Next Level', 24–26 July 2022, which I co-organised with Tim Burns. I am grateful to him for many things, but especially for expertly chairing a public discussion with a significant number of participants from around the world during that seminar on how best to publicise Stein's work for Husserl.

Dirk Fonfara as Editor and Julia Jansen as Director of the Husserl Archives in Leuven graciously made available to the contributors the forthcoming edition of the manuscripts Stein had at her disposal for composing the Marly Biemel edition of *Ideas II* from 1952. These are planned to replace the latter as Husserliana's volume IV.

Henning Peucker generously shared his significant knowledge of Husserl's manuscripts and overall position and development during a co-taught Erasmus+ module on the constitution of the psyche in *Ideas II* and *On the Problem of Empathy* at the University of Paderborn, June 2025. The Concluding Considerations owes much to this collaboration.

Antonio Calcagno, Emanuele Caminada, Margarethe Hackermeier, Kathleen Haney, George Heffernan, Sara Heinämaa and Julia Jansen all contributed papers to the conference from which this volume draws. For a variety of reasons, their presentations did not end up as contributions to this volume, but their perspectives helped clarify controversial issues and continue to form the background against which the result makes sense.

I am finally grateful to James Smith for his linguistic correction and proofreading. Needless to say, I remain responsible for any remaining mistakes.

About This Book

This book presents a collection of articles, most of which stem from papers given at the conference *Stein's and Husserl's Intertwined Itineraries 1916-25 with Focus on Ideas II,* hosted online, during COVID at the Centre for the History of Women Philosophers and Scientists, University of Paderborn, 20–21 May 2021. The aim of the conference was to contribute towards clarifying the respective positions of Stein and Husserl concerning the relationship between person and intersubjectivity, with an emphasis on identifying the extent to which their respective positions inform *Ideas II*. This is discussed here under four headings: 1. the text of *Ideas II* and the collaboration between Stein and Husserl; 2. intersubjectivity and social ontology; 3. the distinction between the adjacent positions of Husserl and Stein, and 4. the possibility of combining transcendental intersubjectivity with philosophical anthropology.

Contents

About the Editor

Mette Lebech is an Associate Professor of Philosophy at Maynooth University, Ireland, where she has taught since 1998. She holds degrees from the universities of Copenhagen, Louvain-la-Neuve and Leuven and has published widely on human dignity and the philosophy of Edith Stein. Her publications include *On the Problem of Human Dignity. A Hermeneutical and Phenomenological Investigation* (Königshausen und Neumann, 2009); *The Philosophy of Edith Stein. From Phenomenology to Metaphysics* (Peter Lang, 2015); *European Sources of Human Dignity. A Commented Anthology of Texts on Human Dignity* (Peter Lang, 2019) and *Value Phenomenology. Taking Account of Edith Stein's Contributions* (Bloomsbury, 2026). She is founding President of the International Association for the Study of the Philosophy of Edith Stein (IASPES) and Series Editor of *Edith Stein Studies*, Bloomsbury. She is a Fellow at the Centre for the History of Women Philosophers and Scientists at the University of Paderborn.

Abbreviations

AHua	Analecta Husserliana
CWES	Collected Works of Edith Stein, ICS Publications
ESCW	Edith Stein. The Complete Works, ICS Publications
ESGA	Edith Stein Gesamtausgabe, Herder
ESW	Edith Steins Werke, Herder
Hua	Husserliana
HuaDok	Husserliana Dokumente
HuaMat	Husserliana Materialien
ICS	Institute of Carmelite Studies
Ideas II	English translation of *Ideen II*
Ideen II	Husserl, E. (1952). *Ideen zu einer reinen Phänomenologie und phänomenologischen Philosophie. Zweites Buch: Phänomenologische Untersuchungen zur Konstitution*, ed. M. Biemel, Hua IV. Martinus Nijhoff. Alternatively: The idea that this book should exist. Hence the intention to replace the 1952 edition with the manuscripts Stein had at her disposal for the composition of the work: Husserl, E. (2025). *Ideen zu einer reinen Phänomenologie und Phänomenologischen Philosophie. Zweites Buch: Phänomenologische Untersuchungen zur Konstitution und Wissenschaftstheorie sowie das Nachwort zu meinen Ideen. Die drei Teilentwurfe und Thematische Ausarbeitungen (1908–1930).* D. Fonfara (Ed.), *Hua* (Vol. IV/V). Springer.
JPPF	*Jahrbuch für Philosophie und phänomenologische Forschung*

Chapter 1
Introduction

Mette Lebech

Abstract Stein attempted already in her doctorate to complement Husserl's work on phenomenology with a painstaking analysis of empathy, including its indispensable role in the constitution of the psycho-physical individual and the person. As Husserl's assistant 1916–18, she edited his *Ideas II*, perhaps introducing into it a view of intersubjectivity adjacent to Husserl's own. When *Ideas II* was finally published after Husserl's death, Husserl's dissatisfaction with her edition was visible in his many additions. The forthcoming edition of *Ideas II* provides an occasion to discuss Stein's and Husserl's respective contributions to the field of the phenomenology of the social to which they both contributed during the years 1916–25. The present work addresses this topic under four headings: 1. the text of *Ideas II* and the collaboration between Stein and Husserl; 2. intersubjectivity and social ontology; 3. the distinction between the adjacent positions of Husserl and Stein; and 4. the possibility of combining transcendental intersubjectivity with philosophical anthropology. The aim is to clarify the respective positions of the two authors on the matters addressed by *Ideas II*.

Keywords Ideas II · Stein · Husserl · Intersubjectivity · Social ontology · Classical phenomenology · Philosophical anthropology · Transcendental phenomenology

It is generally well known that as early as in her doctoral dissertation, Stein attempted to complement Husserl's work on phenomenology with a painstaking analysis of empathy, including its indispensable role in the intersubjective constitution of the psycho-physical individual and the person (Stein, 1917, 1989). It is also well known that, as Husserl's assistant from 1916 to 1918, she edited his *Ideas II*; however, there has been considerable controversy as to whether she introduced into it a view of intersubjectivity which is adjacent to his. Stein's letters to Roman Ingarden report in detail on her complicated experience of working for Husserl on the manuscripts

M. Lebech (✉)
Maynooth University, Maynooth, Ireland
e-mail: mette.lebech@mu.ie

M. Lebech (ed.), *Stein and Husserl's Phenomenology of the Social in Ideas II*, Women in the History of Philosophy and Sciences 25,
https://doi.org/10.1007/978-3-032-17339-3_1

of *Ideas II* (Stein, 2001, 2014a, 2014b), and the dates are chronicled by Schuhmann (1977, pp. 202–222). When *Ideas II* was finally published long after Husserl's death, Husserl's dismay at Stein's redaction shone through in his many additions to the text (Husserl, 1952, 1989). The new edition by Dirk Fonfara of the manuscripts Stein had at her disposal for compiling *Ideas II* (Husserl, forthcoming) provides an occasion to discuss anew Stein's and Husserl's respective contributions to the phenomenology of the social during the years 1916–25.

The present collection of articles aims to further the understanding of the respective positions and intentions of the two authors working on this seminal publication, and to shed light on the relationship between person and intersubjectivity (between personal and intersubjective constitution), which is one of the matters at stake in *Ideas II* and in the intertwined phenomenology of the two authors during the period of its redaction. The understanding of the relationship between person and intersubjectivity is of clear importance to intersubjective phenomenology generally (e.g., Schutz, 1967, 1974, as well as for the traditions of social construction that rely upon it (e.g., Schutz & Luckman, 1973). How exactly our two authors envisaged this relationship is likewise of importance for adequately understanding both the factors that determined their collaborations from 1916 to 1925 and the development of each independently afterward. Tracing the authors' evolution in this period, whether from evidence found within the period in question or from positions known to have been adopted earlier or later, should enable us to better see the two phenomenologists and to better situate their respective positions—for the first time?—in *relation* to one another, as distinct from superimposed on—or mistaken for—one another.

The contributions to this volume are grouped under four headings: 1. the text of *Ideas II* and the collaboration between Stein and Husserl; 2. intersubjectivity and social ontology; 3. the distinction between the adjacent positions of Husserl and Stein; and 4. the possibility of combining transcendental intersubjectivity with philosophical anthropology. These headings mark the consecutive intonation of subsequent consonant themes rather than the naming of mutually exclusive categories. A certain dialectic between the identity and differentiation of Husserl's and Stein's perspectives is expressed in this organisation—starting with the possibility of an identity of doctrine, moving through various complications related to their distinct positions and respective rationales, towards further possible syntheses that it may still be desirable for future phenomenological research to explore.

Most of the contributions to this volume originated as papers given at the conference *Stein's and Husserl's Intertwined Itineraries 1916–25 with Focus on Ideas II,* hosted online at the Centre for the History of Women Philosophers and Scientists, University of Paderborn, 20–21 May 2021. Additional papers share the overall aim of the conference: to contribute towards clarifying the respective positions of Stein and Husserl concerning the relationship between person and intersubjectivity, with an emphasis on identifying the extent to which their respective positions inform *Ideas II*.

Let me therefore, in the following, first discuss the background against which this aim is pursued (1); then sketch out the discussion to which the papers in this volume refer back (2); introduce the present volume as a means to advance the aim (3); and finally, outline the content of the individual contributions (4).

1.1 Background

When *Ideas II* was initially published with support from UNESCO as the fourth volume of *Husserliana* in 1952, its editor Marly Biemel proposed a text that had been redacted in stages. She identified three "*Ausarbeitungen*" (Husserl, 1952, pp. xvi–xx; pp. 397–401): two by Stein and one by Ludwig Landgrebe. The first (1916) was a transcription by Stein of Husserl's manuscripts in Gabelsberger stenography, which Biemel refers to as a "copy". The second redaction by Stein (1918–19), in contrast, contained "significant synthesisation" (Husserl, 1952, p. 400) and "involved much more than merely transcribing" (Husserl, 1989, p. xiii). It drew upon a variety of Husserl's earlier manuscripts to bring them into a structured whole, such that *Ideas II*'s translators into English—R. Rojcewicz and A. Schuwer—referred to the work's unity as "one that has been imposed on it" (Husserl, 1989, p. xi). Stein's second redaction created this unity and contained "*grosso modo*" the text of Biemel's edition, "except for the supplements" (Husserl, 1952, p. xvii). The third redaction (1924–25) was undertaken by Landgrebe in conjunction with Husserl. This one, however, did not involve "independent expansions of the main text" (Husserl, 1952, p. xviii), but added several texts as "supplements"—"in order to preserve as much as possible the unity of exposition" (Husserl, 1989, p. xiii). Biemel's edition relied on Landgrebe's final redaction, as it was "the last and the most complete", while incorporating the stenographic annotations made by Husserl after its completion (Husserl, 1952, p. xviii).

The fact that Husserl did not himself bring the work to publication may indicate that he was not satisfied with it even in its final shape. It nevertheless served in the years after its posthumous publication as an anchor point for the interpretations of Husserl's understanding of intersubjectivity, without explicitly discerning in it the possibly distinct voice of Stein (Hackermeier, 2008, pp. 38–39). It could be argued that the work did that relatively well: Rojcewicz and Schuwer underlines how both Heidegger and Merleau-Ponty took inspiration from it, Heidegger already from before *Being and Time*, and Merleau-Ponty after Husserl's death while working in the Husserl Archives in Leuven (Husserl, 1989,p. xvi). Both Heidegger and Merleau-Ponty would have known first-hand of Stein's collaboration—Heidegger because of his publication of the *Lectures on Inner Time Consciousness* (Husserl, 2013), which had also been edited by Stein, and Merleau-Ponty because he worked on the manuscripts in the Husserl Archives, where Stein's *Nachlass* was also kept at the time. The success did not therefore rely on these authors being unaware of Stein's collaboration. It rather seems to point to the closeness and compatibility of Stein's and Husserl's positions, including in the eyes of the first interpreters.

However, the diversity of author intentions perhaps comes to light once the work is closely and critically examined, as distinct from serving as an inspiration regarding the themes of which it treats (the constitution of nature, psyche and spirit). Sawicki, for example, describes the 1952 text as "troubled" (Sawicki, 1997, p. 73), and as pursuing "two distinct and utterly incompatible strategies to solve the problem that it means to address": strategies which she associates with the diversity of authors.[1] Lohmar refers to the text as "untransparent" and "distorted" , (Caminada, 2018, p. v), whereas Caminada identifies Stein's separate intention as motivated by a desire for "systematisation", which is ultimately directed by her own interest in the constitution of the human person, and not by Husserl's aim , (Caminada, 2018, p. 124, note 2).

The question of how—or whether—to interpret Stein's and Husserl's other contributions to the phenomenology of intersubjectivity in the light of *Ideas II* hangs on the determination of what exactly it was (alternative strategy, development, distortion or synthesis) that Stein contributed as she developed her second redaction of the work. It also hangs on what exactly *Husserl's aim* in it was. To avoid any unhelpful speculation, the Husserl Archives has decided to publish the manuscripts that lay before Stein as she wrote up her second elaboration, in the place of the "final" version (Biemel's) that first brought *Ideas II* to the world (Husserl, forthcoming). This new edition was kindly made available in pre-publication format to the contributors to this volume by its editor Dirk Fonfara.

1.2 Outline of the Discussion

Sawicki's proposal of contrasting strategies present in *Ideas II* , (Sawicki, 1997, pp. 153–162) was taken up by Hackermeier, who read §§ 35–42 of *Ideas II* as being part of Stein's development towards proposing a realist phenomenology to counteract Husserl's transcendental idealism , (Hackermeier, 2008, pp. 84–103). However interesting and groundbreaking Sawicki's proposal is, and however convincing Hackermeier's account of Stein's hidden influence on Continental Philosophy is, their interpretation of the passage in which Stein indicates her own position's divergence from that of Husserl's may be mistaken. Sawicki takes Stein's "points on which the Master and I differ (the necessity of a body for empathy)"[2] to indicate that Stein understands the body to be "a prerequisite for constitution" , (Sawicki, 1997, p. 159).[3] However, "the necessity of a body for empathy" probably refers instead to the fact that Stein

---

[1] At other times, Sawicki seems to attribute the inconsistency to Husserl's own thought: Stein "was supposed to check out Husserl's transcendental phenomenology and vouch for its validity. In order to do that, she had to let her I ride through the moves made in his manuscripts. At many points, Stein found that she was able to re-enliven the 'tending' of Husserl's thought within her own i. But [...] Stein's i-ride through *Ideen II* brought her up against many places where no egoic validation was possible because Husserl's argument ran in circles." , (Sawicki, 1997, p. 226).

[2] Letter to Ingarden 20 March 1917 (Stein, 2001, p. 51).

[3] This possibly puts an emphasis on Stein's "realism" that may not be warranted (even if the "realism" that would interest Sawicki would be a feminist one, the purpose of which would be to vindicate, against certain types of feminism, science as not inherently oppressive; Sawicki, 1997, 250 ff.). Although this would reflect a genuine Steinian concern, Stein would not have formulated it in terms of "realism". Another letter to Ingarden in fact testifies to Stein's (re-) conversion to idealism

did *not* consider a body to be essentially required for empathy, whereas Husserl *did*. In the final paragraph of *On the Problem of Empathy*, Stein emphatically states that it is *unclear* whether it is possible to empathise "with purely spiritual persons" (Stein, 1989, p. 118; 1917, p. 131) and the structure and intent of her *Contributions* depends on the possibility of distinguishing psyche from spirit. Husserl, in contrast, brings back into *Ideas II* the sentence (which had been discarded by Stein) of Supplement XII, I: "The spirit has a psychic basis". He is thereby in principle affirming that spirit *as such* is psychic in nature (Husserl, 1989, p. 345; 1952, p. 333), consonant with his idea that time-consciousness pertains to subjectivity as such.[4] That Stein mentions her divergence from Husserl means that she must have discussed it with him, likely in detail, to know for certain that their points of view diverged on this matter. This points to the possibility that Stein may have felt unable to follow Husserl's thinking and to complete *Ideas II* in accordance with it, because she thought his intuition on this matter was insufficiently clarified.

The two incompatible strategies that Sawicki sees at work in *Ideas II* are described as two solutions to the problem of "establishing a single foundation for the natural and cultural sciences". One solution, she claims, "founds both kinds of science in the immediate intuitions available to living animal bodies such as our own". The other solution, in her words, "founds the sciences in the activity of object-constitution carried out by transcendental subjectivity" , (Sawicki, 1997, pp. 73–74). For both Stein and Husserl, however, these two strategies are not incompatible *in so far as science is limited to human (as distinct from divine) science*, since constitution is the identification of what is available to "us" [human beings] in immediate intuition, inclusive of data that testifies to "us" [human beings] having bodies and psyches.[5] To Stein, indeed, the body and the psyche are clearly constituted from "our" experience, but it is imaginable that "I"s could have very different kinds of bodies and psyches, or indeed have none at all. In contrast, this latter possibility does not seem to be considered by Husserl. To him, a subject seems to be *invariably* psychic and thus linked to a body through which it is indispensably identified.[6]

(24 June, 1918: Stein, 2001, p. 87; 2006, p. 110). In *Finite and Eternal Being*, Stein distinguishes between ideal and real being to overcome the fruitless dichotomy (Stein, 2006, 62 ff.).

[4] This idea is discussed and denied by Stein in *Finite and Eternal Being*, § 6 (Stein, 2006, 51 ff.).

[5] It is worth quoting Stein's June 1918 letter: "N.b.: I have turned over a new leaf when it comes to idealism and believe it can be understood in a way that is metaphysically satisfying. It appears to me, however, that much of what is presently in the *Ideen* has to be composed differently, though in Husserl's sense, if only he brings together what has, and in a critical moment does not leave out of consideration something that necessarily belongs to the subject. (I think, for example, of the recasting of the concept of consciousness that results from a consideration of the constituting of consciousness – something you also wish). [...] Of course the discussions are beautiful and valuable, but they prevent us from ever achieving a synthesis." (Stein, 2006, pp. 110–111). If Stein's challenge was to achieve a synthesis on behalf of Husserl, the question of what was involved is crucial, since it is not obvious whether Husserl shared it.

[6] Our interpretation of Stein's description of her disagreement with Husserl is not completely irreconcilable with Sawicki's and Hackermeier's. One could explain that Stein's description of the constitution of the body seems more sustained than Husserl's by her ability to discern in experience the specificity of *having* body and psyche, due to the latter not being confused with spirit.

There seems to be agreement among scholars that Stein's redaction of *Ideas II* is at its most self-directed in the second half of Section II (*The Constitution of Animal Nature*, containing Chapters 3–4: "The Constitution of Psychic Reality Through the Body" and "The Constitution of Psychic Reality in Empathy"). Here, according to Rojcewicz and Schuwer, Stein "go[es] back to numerous manuscripts written by Husserl during the war years, up to the beginning of 1917" (Husserl, 1989, p. xii), and according to Biemel she "put a new text together".[7] Sawicki refers to Chapter 4 (§§ 43–47) as "the hinge chapter between 'nature' and 'culture'" and states that "[a]pparently, Stein herself composed this" , (Sawicki, 1997, p. 158). It is in these chapters, which Sawicki classifies as "anonymous textual production" and likens to "ventriloquism" ,(Sawicki, 1997, p. 151; p. 168), that Stein's perspective most clearly "distorts" Husserl's, or conceives the unity of "his" position according to hers.[8] It makes sense that Stein's distinctive subjectivity would ventriloquise through the patchwork of Husserl texts that she composes in Chapter 3 and 4, such that a schema already present in her *On the Problem of Empathy* is "imposed" on the text of *Ideas II*.

Did Stein intend to "rectify" Husserl's thought? Among Stein's intentions in writing anonymously for Husserl, Sawicki is probably right to point out that "Stein employed [the strategy of offering her services as assistant] to get access to the thought of Husserl for herself and to ensure that it would be available for others". Stein makes a point later in her *Beiträge* of explicating in detail the distinction between psyche and spirit, such that these can be read as a kind of parallel to Husserl's later "Supplements" to *Ideas II*: Similarly to how Husserl adds what he feels is missing to see his own intuitions adequately represented in *Ideas II*, so Stein provides her *Beiträge*, to add what she feels is missing in the work to see her intuitions adequately represented.[9]

The development of genetic phenomenology, which Lohmar finds already present in the manuscripts for *Ideas II* (Caminada, 2018, p. 1), might have seemed to Stein to be incompatible with a commitment to eidetic phenomenology, in so far as it depends on an undifferentiated psyche-spirit conception. To Stein, eidetic imagination was the last recourse for phenomenology—not habit, however entrenched. Habit, she thought, could only be discerned by someone to whom eidetic intuition

---

[7] „[…] stellte den Text […] ganz neu zusammen"(Husserl, 1952, p. xvii).

[8] Stein's nickname among her friends in the Bad Bergzabern Circle, "Anonyma," probably plays on the personality trait that made such work possible—not without revealing a certain ironic distance to it at the same time (Sawicki, (Feldes, 2021;Sawicki, 1997, p. 152). Sawicki's reflections on this type of work reflects the same distance: "Phenomenologically, we may say that the anonymous writer purports to write as if she were merely reading. (She pretends that) someone else's i has 'originarily' lived the experiences that are coming to expression on the paper in the act of writing, while her own i is merely following along 'nonoriginarily' when it writes. The emergent expressions supposedly owe nothing to her, but are entirely determined by the original meaning-enactments of someone else" (ibid.). Sawicki is of the opinion that "Husserl's expectation was that Stein would work in this manner." She also is adamant that "[a]nonymity as a strategy of writing, does not succeed in its avowed purpose. The writer (like the reader) of course cannot help creating a version of the material in which her own distinctive subjectivity is a decisive ingredient."

[9] Please see Chapter 12, "Concluding Observations", for an elaboration of this.

was already available, and thus genetic phenomenology may have looked to her like a naturalisation of phenomenology, bound to short-circuit the entire enterprise.

That *Ideas II* should concern "constitutional matters" in the first place was not explicitly stated in the announcement of the work in *Ideas I*.[10] However, the focus on constitution became obvious from about 1912, as it became one of the main problems of phenomenology (Husserl, 1952, p. xv). With her thesis *On the Problem of Empathy*, Stein had obtained Husserl's respect and approval for distinguishing empathy's contribution to the constitution of the psycho-physical individual (Chapter 3) from its contribution to the constitution of the person (Chapter 4). A division of the manuscript of *Ideas II* into three parts dedicated to the constitution of material nature, animal nature and the spiritual world thus seemed apposite, and cannot have met with much resistance from Husserl. Husserl had announced an analysis of how the physical sciences, psychology and the cultural sciences relate to phenomenology (Husserl, 1983, p. xxi), and what could accomplish that *but* a constitutional analysis of nature, psyche, and spirit?

Yet, as Stein seems to have found herself left alone to bring the manuscripts for *Ideas II* into a publishable whole, she took the constitution of the psycho-physical individual and of the spiritual person to mean *what she meant by them* in her own work (Stein, 1917, 1989). In so far as this involves—as indicated above—no recognition of an eidetic necessity of a psychic foundation for the person, she would have conceived of the role of the body in intersubjective communication accordingly. That Stein could imagine a purely spiritual person, and that the experience of such a person's experience could have importance for my constitution of myself and of others, seems confirmed by Stein's later work. That Husserl did not consider this possibility and instead took the spirit to be essentially linked to the psyche, is confirmed by Dieter Lohmar's interpretation of Husserl to regard the body as the "*indispensable* medium of the common constitution founded in mutual understanding" [my emphasis].[11]

---

[10] "In the *Second Book* we shall then treat in detail some particularly significant groups of problems, the systematic formulation and characteristic solution of which are the precondition for being able to make actually clear the difficult relationships of phenomenology, on the one hand, to the physical sciences of Nature, to psychology and to the cultural sciences; on the other hand, however, to all the a priori sciences" (Husserl, 1983, p. xxi, <5>).

[11] „Der Leib ist unentbehrliches Medium der im gegenseitigen Verstehen begründeten gemeinsamen Konstitution", (Caminada 2018, p. vii). He continues: „Das gegenseitige Verstehen und die gemeinschaftliche Konstitution einer geistigen Umwelt in der Gemeinschaft hat dann als *Voraussetzung* und *Grundlage* eine *jederzeit* erlebte und *anschauliche* Leiblichkeit, die ich mit *anderen gemeinsam* habe und die mich mit ihnen durch Verstehen verbindet [my emphasis]". The added emphasis refers to where Stein is likely to have diverged slightly in her understanding, according to the interpretation proposed here: to her, in so far as she does not consider empathy to be necessarily reserved for bodily individuals, a subtle "Erweiterung" of the proposed would have to be envisaged. Lohmar continues: „Dem Leib kommt in dem neu rekonstruierten Gefüge der konstitutiven Schichten eine zentrale Rolle zu. Er wird als ein *unmittelbar anschauliches Erfahrungsfeld* betrachtet, das mit meinen Bewusstseinserlebnissen *direkt* zusammenhängt [my emphasis]."Stein, by contrast, is interested in the *constitution* of the body, which may be seen as something specific only when the possibility is contemplated that *no* body be constituted in (someone's) experience, including my own. She is likely to have regarded it as impossible to fully constitute the body if the possibility of not constituting it would not simultaneously present itself to the imagination. The

These differences occasion Stein's subsequent *Contributions to the Philosophical Foundation of Psychology and the Humanities* (Stein, 2000, 2010).[12] Stein hints at her (lack of) contribution to *Ideas II* when she refers to the anchor point of her *Contributions* as being not the long-awaited and at the time unpublished *Ideas II*, but her own *On the Problem of Empathy*.[13] While acknowledging her profound debt to Husserl's thought,[14] she may all the same have worked very hard to let his perspective stand in *Ideas II*, despite her doubts about its adequacy. It seems possible to see this discrete but consequential difference as ultimately guiding the two authors along different paths.

The correct diagnosis of the divergent positions is of importance for Stein studies and Husserl scholarship alike. It concerns a crucial moment in the development of both authors' thought about what exactly intersubjectivity and empathy contributes to the constitution of psychic and personal individuals. Very likely, Fonfara's introduction to the manuscripts of *Ideas II* will also address this question. All the contributions in this volume attempt, in different ways, to identify the respective trajectories of the two authors and to discuss how they combine or intersect in the drafting of *Ideas II*.

---

indispensability of the body to *any* personal experience is therefore, to her, mistaken. This should be apparent to Husserl as well, since he too distinguishes between the natural layer of the soul and the spiritual person, from the point of view of the "personal attitude" , (Caminada 2018, p. ix). It is arguably this inconsistency which Stein attempts to "clear up", without success. It would possibly amount, in her view, to a confusion of the respective objects of psychology (the psyche) and the humanities (the spirit). That the eidetics of the spirit would be called "rational psychology" compounds this confusion (x).

[12] "The following investigations undertake to penetrate into the essence of sentient [*psychischen*] reality and of the mind [*des Geistes*] from various sides, and thereby to secure the groundwork for a definition of psychology and the humanities that will fit the facts" (Stein, 2000, p. 1; 2010, p. 3). If Stein thought that complete constitutional analyses of the psyche and the spirit had already been accomplished in the manuscript she had completed for Husserl, it would not have made sense to "contribute" these, nor to explicitly refer to her work as "contributions" to this task.

[13] "The problems whose solution is attempted here already surfaced in my dissertation, 'On the Problem of Empathy'. In the context of analysing how you undergo another subject's experiences, I felt I needed to sketch the composition of the human personality in broad strokes without being able in that format to tackle an in-depth investigation of the complicated issues surrounding this problem. The first of the two following investigations *Psychic Causality*] attempts that now: to work out plainly the twofold basic lawfulness – causality and motivation – operating together within one sentient subject with a sensuous-mental essence. The second investigation [*Individual and Community*] broadens the consideration from the isolated sentient individual to super-individual realities, and thereby seeks to attain broader insights into the composition of the mental cosmos" (Stein, 2000, p. 1; 2010, p. 3).

[14] "[I]mportant influences on my own work came out of the stimulation that I was receiving [... to the point that] very often I was not sure whether I would have to regard something as my own research result or as an internal appropriation of transferred thought motifs" (Stein, 2000, p. 2).

1.3 The Present Volume

The discussion of *Ideas II* is important not only for an accurate assessment of the work of both philosophers, but also to aid the advancement of the field of social phenomenology as such, which has great potential for supporting the disciplines of the social sciences and politics. Thus, the discussion Stein expected from Husserl of issues fundamental to phenomenology as such—the discussion of how psychic causality and practical eidetics mutually condition each other—may be one that pertains to the sociology of knowledge which we still have to accomplish. Could Husserl's unwillingness to engage with Stein's revisions have stemmed from him simply not seeing her point? Everyone advances in their own direction, and the discussion *is* daunting.

Distinguishing the respective contributions of the two philosophers to *Ideas II* allows us to respect and assess the intrinsic integrity of the work of either. As long as one contribution is presumed to be subsumed under or in the other, the resulting amalgamation will not contribute towards an understanding of the proper development of either author, or indeed of their respective concerns. Whereas a synthesis of both positions may show itself to be a fruitful way to understand social phenomenology—the contemporary notion of "phenomenological empathy" unhesitatingly fuses both contributions—knowing what distinguishes fusion and synthesis may be necessary for the synthesis to function in the long run.

Knowing this necessitates a refinement of current positions as well as greater nuance and clarity in the phenomenological treatments of social reality. Since the phenomenology of intersubjectivity and the social have hitherto relied largely on interpreting *Ideas II* as the work of one person, the clarification of the distinct positions at work in this work can help to clarify the field and create a more secure foundation for subsequent developments in two ways. The first is by highlighting and including Steinian insights in the tradition issuing from *Ideas II* and following them through in an internal updating of this tradition. The second is by insisting on them constituting an alternative voice—one already breaking out from within the tradition, but now providing a surreptitious critical distance to it, since it needs to be seen as the work of another author and therefore in the context of this other author's work. The differentiation of positions—or the adoption of a more conscious synthesis of the distinct positions—should allow phenomenology to provide better analyses in the areas of intersubjectivity and to map the possible and factual types of personal constitution of intersubjectivity of relevance to socially constructed reality, and in this way to fulfil its promise for social science. The phenomenology of the social is increasingly influential in social science and has great potential to support this discipline. It has that potential to the exact extent that phenomenology fulfils its promise of providing a philosophical foundation by clarifying the fundamental concepts of that science phenomenologically.

1.4 The Individual Contributions

The collection is opened by two contributions which discuss the text of *Ideas II* and the collaboration between Stein and Husserl. First, Anna Varga-Jani discusses the working relationship of the two authors, in "The Collaboration Between Husserl and Stein. Historical and Phenomenological Approaches". She does so by identifying similarities and differences in their respective understanding and use of phenomenology and by carefully tracking Stein's development compared to that of "the Master". Fr Francis B. Payo proposes—in "Husserl and Stein on Empathy: Paving a Way Towards Integration"—to integrate Stein's account of empathy with Husserl's, given the difficulty there is in separating them from each other. An integration of their respective understandings would be a presupposition for regarding *Ideas II* as having a coherent author, and as it has often been read as such, its rationale ought to open the discussion.

The second section discusses intersubjectivity and social ontology in three papers. These all highlight how differences make manifest the distinct voices of the two authors, despite their close collaboration in establishing the text of *Ideas II*.

Francesca De Vecchi—in "Personally Embedded Qualitative Social Ontology: Issues from Stein and Husserl"—points out some of the differences between Husserl's and Stein's perspectives on the social, but also show how both authors focus on the relationship between personal identity and personal life, on the one hand, and between personal collectives and social life, on the other. Pól O'Hagan's "Us & Them: Stein/Adorno and Socially Constituted Identity" discusses how some key tenets of Theodor Adorno's critique of Husserl bear a striking similarity to positions advanced by Stein, whose understanding of intersubjectivity addresses the entrenched individualism which Adorno found so problematic in the work of Husserl and Heidegger. Engagement with Stein through Adorno (and with Adorno through Stein) opens up, for O'Hagan, the potential for "an innovative re-alignment of Frankfurt School Critical Theory" in phenomenological terms. James McGuirk discusses "Empathy in the teaching/learning relation" in relation to Stein, Husserl, and Hannah Arendt. He argues that the phenomenological analyses of empathy offered by Husserl and Stein are useful for understanding that relationship, but that Stein's account, putting more emphasis on the experience of *foreign* consciousness, provides a richer description of intersubjective world-constitution (resisting the interchangeability in principle of conscious perspectives on the world), which brings it closer to Arendt's discussion of the teaching/learning relationship.

The third section takes stock of the many themes so far laid out and attempts to discern the distinct positions of Husserl and Stein in three chapters.

Valentina Gaudiano's "Stein's Concepts of 'Life power' and 'Core' in the Description of the Living Body" claims that the phenomena of "life power" and "core" are the phenomena which distinguish Stein's account of the foreign and own living body from Husserl's. Gaudiano treats of these in the context of the many connections between the two thinkers, while at the same time highlighting Stein's originality. Carlos Guillermo Viaño Rubio's "Edith Stein and Edmund Husserl on Phenomenology. The Balance of the Steinian Reception of Transcendental Phenomenology in *Introduction to Philosophy*" assesses the difference between the perspectives of Stein and Husserl from an analysis of Stein's *Introduction to Philosophy* (1921) and its discussion of (Husserl's) transcendental idealism. Next, in my article—"Stein's Contribution to the Phenomenological Foundation of the Social Sciences in Conversation with Schutz's"—I discuss Stein's contribution to the phenomenological foundation of the social sciences in a conversation with Alfred Schutz, whose account of the same relied on Husserl's phenomenology of intersubjectivity. A core text for Schutz is *Ideas II*, and—should Stein's critique of that work in her *Beiträge* be taken into account—a combination of Stein's and Husserl's perspectives could, as is argued, place sociology on transcendental foundations.

This sets the scene for the collection's fourth section, which discusses the relationship between transcendental intersubjectivity and philosophical anthropology in two papers. Mariéle Wulf's "Stein's Philosophical Anthropology as a completion of Husserl's Project for a Foundation of the Sciences. *Ideen II* – The 'Master' and His Co-Genius" describes Stein's philosophical anthropology as a completion of Husserl's project for a foundation of the sciences. Wulf emphasises how the quality of the personal I influences the vision of the object, both subjectively and objectively, thus bringing the discussion of how personal and intersubjective constitution intersect in the respective views of Husserl and Stein to a (preliminary) close. Angela Ales Bello proposes, in her "Rethinking Edmund Husserl's and Edith Stein's Philosophical Anthropology and Metaphysics", an analysis of the relationship between Husserl's and Stein's philosophical anthropology and metaphysics. Her study underscores phenomenology's specific approach to the human being, the world, and God, which Husserl and Stein share. She discerns a greater similarity in philosophical anthropology than in metaphysics, and emphasises how Stein's turn to Scholasticism in metaphysics distinguishes it from Husserl's metaphysics (upon which she nevertheless insists).

My final "Concluding Observations" read Stein's *On the Problem of Empathy* as testifying to the presuppositions with which she started her work on *Ideas II*, and her *Contributions to the Philosophical Foundation of Psychology and the Humanities* as stating her critique of what she took Husserl's position to be in the work she put together for him. It argues that *On the Problem of Empathy* provides the rationale for the structure of the 1952 edition of *Ideas II*, and that *Contributions* explains the reasons behind that structure in the form of a critique of the result.

***

Future research will no doubt focus on developing intersubjective phenomenology, relying on both Husserlian and Steinian contributions in different proportions and for different reasons. It is likely, however, that further discussion of the distinct purposes and objectives of the two thinkers will be needed to support these developments. Hopefully, the contributions contained in the present volume can serve as a starting point.

References

Caminada, E. (2018). *Vom Gemeingeist zum Habitus: Husserl's Ideen II. Sozialphilosophische Implikationen der Phänomenologie*. Springer.

Feldes, J. (2021). "Anonyma" und ihre geistigen Lebenspole. Zur Entwicklung von Edith Steins Christlicher Philosophie in ihren sogenannten "Speyerer Jahren". In H. Klueting & E. Klueting (Eds.), *Edith Steins intellektueller Weg* (pp. 3–14). Aschendorf.

Hackermeier, M. (2008). Einfühlung und Leiblichkeit als Voraussetzung für intersubjektive Konstitution. In *Zum Begriff der Einfühlung bei Edith Stein und seine Rezeption durch Edmund Husserl, Max Scheler, Martin Heidegger, Maurice Merleau-Ponty and Bernhard Waldenfels*. Verlag Dr Kovač.

Husserl, E. (1952). *Ideen zu einer reinen Phänomenologie und Phänomenologischen Philosophie, II. Hua IV* (M. Biemel, Ed.). Martinus Nijhoff.

Husserl, E. (1983). *Ideas pertaining to a pure phenomenology and to a phenomenological philosophy* (I. F. Kersten, Trans.). Kluwer.

Husserl, E. (1989). *Ideas pertaining to a pure phenomenology and to a phenomenological philosophy II* (R. Rojcewicz & A. Schuwer, Trans.). Kluwer.

Husserl, E. (2013). *Zur Phänomenologie des inneren Zeitbewußtseins*. (In R. Bernet, Ed.). Meiner.

Husserl, E. (forthcoming). Ideen zu einer reinen Phänomenologie und Phänomenologischen Philosophie. Zweites Buch: Phänomenologische Untersuchungen zur Konstitution und Wissenschaftstheorie. In D. Fonfara (Ed.), *Die drei Teilentwurfe und Thematische Ausarbeitungen sowie das "Nachwort meiner Ideen" (1908–1930), Hua* (Vol. IV/V). Springer.

Sawicki, M. (1997). *Body, text and science. The literacy of investigative practices and the phenomenology of Edith Stein*. Kluwer.

Schuhmann, K. (1977). *Husserl-Chronik. Denk- und Lebensweg Edmund Husserl's*. Martinus Nijhoff.

Schutz, A. (1967). *The phenomenology of the social world*. Northwestern University Press.

Schütz, A. (1974). *Der sinnhafte Aufbau der sozialen Welt. Eine Einleitung in die verstehende Soziologie*. Suhrkamp.

Schutz, A., & Luckman, T. (1973). *The structures of the life-world*. Northwestern University Press.

Stein, E. (1917). *Zum Problem der Einfühlung*. Buchdruckerei des Waisenhauses.

Stein, E. (1989). *On the problem of empathy CWES 3* (W. Stein, Trans.). ICS Publications.

Stein, E. (2000). *Philosophy of psychology and the humanities. CWES 7* (M. C. Baseheart & M. Sawicki, Trans.). ICS Publications.

Stein, E. (2001). *Selbstbildnis in Briefen III. Briefe an Roman Ingarden, ESGA 4* (H.-B. Gerl-Falkovitz, M. A. Neyer & E. Avé-Lallemant, Eds.). Herder.

Stein, E. (2006). *Endliches und ewiges Sein. Versuch eines Austiegs zum Sinn des Seins, ESGA 11–12* (A. U. Müller, Ed.). Herder.

Stein, E. (2010). *Beiträge zur philosophischen Begründung der Psychologie und der Geisteswissenschaften, ESGA 6* (B. Beckmann-Zöller, Ed.). Herder.

Stein, E. (2014a). *Letters to Roman ingarden. CWES 12* (H. Hunt, Trans.). ICS Publications.

Stein, E. (2014b). *Freiheit und Gnade"und weitere Beiträge zu Phänomenologie und Ontologie (1917–1937)* (B. Beckmann-Zöller & H. R. Sepp, Eds.). Herder.

Mette Lebech is Associate Professor of Philosophy at Maynooth University, Ireland, where she has taught since 1998. She holds degrees from the universities of Copenhagen, Louvain-la-Neuve and Leuven. She has published widely on human dignity and the philosophy of Edith Stein. Her publications include *On the Problem of Human Dignity. A Hermeneutical and Phenomenological Investigation* (Königshausen und Neumann, 2009); *The Philosophy of Edith Stein. From Phenomenology to Metaphysics* (Peter Lang, 2015); *European Sources of Human Dignity. A Commented Anthology of Texts on Human Dignity* (Peter Lang, 2019) and *Value Phenomenology.*

Taking Account of Edith Stein's Contributions (Bloomsbury, 2026). She is founding President of the International Association for the Study of the Philosophy of Edith Stein (IASPES), Series Editor of *Edith Stein Studies,* Bloomsbury and Fellow at the Centre for the History of Women Philosophers and Scientists at the University of Paderborn.

Part I
The Text of *Ideas II* and the Collaboration Between Stein and Husserl

Chapter 2
The Collaboration of Husserl and Stein. Historical and Phenomenological Approaches

Anna Varga-Jani

Abstract Although the relationship between Edmund Husserl and Edith Stein is well documented in phenomenological research, Stein's particular contributions and her changing attitude to Husserlian phenomenology are less frequently appraised and reconstructed into the gradual steps that led her to her own phenomenological concepts. In what follows, I would like to contribute to filling this gap with a few historical facts and a description of the phenomenological perspective of Edith Stein. The aim of my paper is to shed light on the points of commonality and divergence between Husserl and Stein by reference to the historical facts, and to show the point where these two thinkers divide from each other. In the first part of this paper, I will retrace the common period between Husserl and Stein with some reflections on their individual objectives, while in the second part I will give further reflections on Edith Stein's thinking and on her individual contribution to phenomenological thought in general.

Keywords Empathy · Constitution · Individual · Community · Transcendentality · *Philosophia perennis*

2.1 The Historical Background of the Collaboration Between Husserl and Stein

Though Edith Stein began her philosophical studies in her hometown Breslau, she was introduced to Husserlian philosophy during this early period, and soon decided to continue her studies in Göttingen. Thus, the first phase of the relationship between Husserl and Stein includes Stein's studies in Göttingen, her first decision in favour of Husserl and thus the contemporary school of thought of philosophy, between 1913

A. Varga-Jani (✉)
Collegium Professorum Hungarorum, Makovecz Campus Foundation in Piliscsaba, Piliscsaba, Hungary
e-mail: anna.varga.jani@makovecz-campus.hu

M. Lebech (ed.), *Stein and Husserl's Phenomenology of the Social in Ideas II*, Women in the History of Philosophy and Sciences 25,
https://doi.org/10.1007/978-3-032-17339-3_2

and 1918. After the first semester in Göttingen, influenced by Husserl, Stein made two important commitments to Husserl's phenomenology. The first of these was to write her dissertation about empathy under the supervision of Husserl. The preparation of her thesis on empathy, which was later revised in her dissertation work, was nevertheless mainly independent work on her part, without any clear oversight by Husserl, a fact which left Stein slightly disappointed with Husserl.[1] This first disappointment was followed by the publication of the *Ideas for a pure phenomenology and phenomenological philosophy* immediately after Stein's arrival in Göttingen, which induced her second decision to follow Husserl's thinking. Though she wrote in her autobiography that the *Ideas* "included some reflections which sounded as though their Master wished to return to idealism" (Stein 2016, p. 250), *Ideas I* inspired Stein to a deeper interest in Husserl's thought,[2] and she consequently followed Husserl to Freiburg, where she finalized her dissertation thesis. While defending her dissertation on August 3, 1916, she accepted Husserl's invitation for the assistantship and decided to stay in Freiburg and continue to work as Husserl's assistant (Stein 1993, p. 1).

During her one-and-half-year assistantship, Stein worked with manuscripts of Husserl's which would not appear in publication until many years later. As Husserl's assistant, it was her job to organize these manuscripts systematically, to elaborate them, and to put them together thematically. This method, which was otherwise accepted by Husserl, resulted in an ambivalent evaluation of Stein's work by subsequent phenomenological research. On the one hand, it is recognized that Husserl's manuscripts were saved by virtue of Stein's involvement; on the other, Stein is accused of having hindered the development of the manuscripts due to the transformations, shortenings, and simplifications she carried out on them (Husserl 1952, p. XVI). By contrast, Stein's own understanding provides a rational answer to the allegations, insofar as her aim was to popularize Husserlian phenomenology by her editing of the manuscripts and by preparing them for publication. The method she adopted focused on the thematic standardization and completion of the manuscripts. Based on *Husserl-Chronology*, compiled by Karl Schumann , (Schumann,1977), and the editorial forewords of the Husserliana series, Stein's contributions can be traced in the manuscripts published in Husserliana: in Hua IV–V the *Ideas II–III* (Husserl 1952, 1971); in Hua X, *Zur Phänomenologie des inneren Zeitbewusstseins* (Husserl, 1969); Husserliana Materialien V. *Urteilstheorie. Vorlesung* 1905 (Husserl,

[1] Stein (2016, p. 269): "However, almost immediately, I was given another bitter pill to swallow: he required that, as format for the dissertation, I use that of an analytical dialogue with Theodore Lipps." Thus, in the winter semester of 1914, Edith Stein made the first studies on Lipps and showed Husserl her first drafts. As has become known from Stein's autobiography and from Karl Schumann's introduction to Husserl's excerpt of Stein's thesis, Husserl first dealt with the meaning of empathy in the 1913/14 lecture "Nature and Spirit" (Schumann, 1991, p. 687).

[2] Ibid, p. 250: "...which is not to be confused with the transcendental idealism of the Kantian schools..." Stein spent more than ten years with the transcendental idealism of Husserl and exchanged many letters with Ingarden about this topic. Her development in thinking becomes clear by these exchanges. Cf. Stein (2014a, Nr. 6, 9, 14, 37, p. 111; 2014d, pp. 159–62; 1998, pp. 235–236). Cf. also: Husserl's interpretation of the idealism–realism debate in a letter to Daniel Feuling in April 30, 1933 (Husserl, 1994,p. 88). For more about this topic cf. Jani (2015, pp. 79–109).

2002); Husserliana XVI, *Ding und Raum* Ms.: M III 3 V. *Systematische Raumkonstitution* (Husserl, 1973a); Hua XXV, *Aufsätze und Vorträge: Phänomenologie und Erkenntnistheorie*; cooperation in Hua XXXIII, *Die Bernauer Manuskripte über das Zeitbewusstsein* (Husserl, 2001). Husserl's special influence manifests in the dissertation thesis of Stein too, on the topic of the phenomenology of intersubjectivity, although Stein never reflects on these manuscripts (Husserliana XIII) of Husserl's.

Stein's first work for Husserl focused on the organization of the manuscript of *Ideas II*. In 1915, Husserl had undertaken a new elaboration of these texts, incorporating additions from the manuscript of 1912 and from his lecture notes for the seminars of 1913 and 1915 (Husserl, 1952, pp. XVI–VII). Stein described her work in a letter to Roman Ingarden in the following way:

Regarding the nature of my work on *Ideen*, you are probably still not completely clear: for the time being, working only with the material at hand, I am attempting to lay out in a unified draft the entire course of thought (that seems rather apparent but is nowhere fixed or worked out). That is to be the basis of the Master's work, and I really would like to finish it because I believe that he would never find his way through the material and would always get stuck on individual points. (Stein, 2014a, p. 39).

Immediately after completing the *Ideas* during Husserl's summer holiday stay in Bernau, Stein began to work on the topic of space-constitution. In 1907, and especially in the lectures of the summer semester in 1907, Husserl investigated the problem of the constitution of space in close connection with that of the constitution of things. The relevant manuscript (F I 13) also served as the main text of an elaboration produced by Edith Stein in 1917, which Husserl classified under the headline "Systematic Space-Constitution" (Husserl, 1969, p. XIV, n.1).

In July 1917, Edith Stein began to create a certain unity in Husserl's notes concerning time consciousness, and according to her letter to Roman Ingarden, she came across Husserl's lectures on that topic from 1903.[3] These early texts on the phenomenology of time by Husserl, which were published by Martin Heidegger in the *Yearbook for Philosophy and Phenomenological Research* Vol. IX under the title "First Part: The lectures on the inner time consciousness from the year 1905" (Husserl, 1928), took their starting point from the description of the perception of temporal objects (continuous tone) or objects with temporality (duration of the tone). During his stay in Bernau, Husserl edited the part of the lecture from 1904/05 dedicated to the awareness of time. On this occasion, he made extensive changes to the text, and, following the resumption of the problems of inner time awareness, he wrote the so-called Bernauer manuscripts. Edith Stein herself reported on these matters in a letter to Ingarden: "This afternoon I want go down to Husserl who has been in Bernau for eight days. […] Last month, I worked on Husserl's notes on time. They are really beautiful but not yet fully developed" (Stein, 2014a, p. 76).

[3] Cf. Stein (2014a, p. 73): "Now, I have come to a bundle labeled 'time consciousness'. You know best how important this material is for the theory of constitution, for the discussion with Bergson, and it seems to me, for others, for example, Natorp."

Stein regularly informed her friends about her current work. Sometimes, and increasingly as time went by, these reports in letters displayed her growing dissatisfaction and impatience with Husserl. Her expectations of Husserl, that he should find more time to exchange views and ideas during her time as assistant, were more and more unsatisfied. Without any advice from Husserl, she found her work on the multitude of manuscripts even more hopeless, and lost sight of the unifying system according to which she wanted to work.

Presently, he [Husserl] is concerned with the constitution of nature (of course without consideration of the draft[4]). In the meantime, on my own authority and without encountering any opposition, I continue to work on the draft, and I have made it to "person".[5] Of course, this means we hardly still speak to one another. For me, that is very painful, for the issues are very complicated and the material is quite incomplete. Thus, at the moment, I work rather independently, and indeed that is rather gratifying, though some exchange of ideas would be quite beneficial. (Stein, 2014a, p. 33).

On February 19, 1918, Edith Stein quit her assistant position with Husserl. One can conclude from her letters to her friends that Husserl did not even think that the two of them were really collaborating on the manuscripts. It was probably not clear to he himself how he wanted to revise the drafts, if at all. Edith Stein reported to her friends about her dejection and clearly identified the problem of working with Husserl:

When the Master recently favored me with a list of instructions for the treatment of his manuscripts (in a friendly manner, of course, but at the same time, I cannot tolerate something like that), I explained (of course, also in a friendly manner) that the arrangement (1) basically, is impossible; (2) in general, insofar as it could be done, it could be best be done by him, for himself; (3) I would be especially unsuited for it and could endure it all only if I were able, in addition, to do some of my own work. I am curious what he will say about that. I offered to remain in Freiburg and help him with the editing of the *Jahrbuch* and some other work but not as his assistant on work whose meaning is not clear to me. (Stein, 2014a, p. 88)

It is just as difficult to judge Husserl's philosophical influence on Edith Stein as it is to carry out an adequate evaluation of Stein's assistantship. As we can see, Edith Stein was closely connected with Husserl's phenomenology from 1913 to 1918, which is also noticeably reflected in her early works. But even at this time one cannot talk about a real exchange of views between them. Besides the difference in working methods which makes the comparison of their thinking impossible, their common philosophical interest can be limited to terms and phenomenological topics, which led the two philosophers to different directions of thinking. When Edith Stein discussed such phenomenological problems in her early contributions that she developed under influence by Husserl, it becomes obvious that it was in Husserl's phenomenology that

[4] According to the translator of the letters, she means the draft of the first section of *Ideas II*, in the earlier edition, cf. Husserl (1952) and Stein (2014a, p. 35, n. 8).

[5] Cf. Stein (2014a, p. 35, n. 9). See the third section of *Ideas II* (Husserl, 1952) which corresponds to "Teilentwurf 2" in the new edition of *Ideas II* (Husserl, 2025). Cf. Jani (2015, pp. 30–31).

she found the starting point for her own philosophical path. Moreover, the analyses of Edith Stein's phenomenological approaches, in comparison with Husserl, indicate that in almost every case she refers to Husserl's early writings. This also underlines the thesis that rather temporary impulses of Husserlian thinking can be documented in her thought, and that a real co-thinking with Husserl never existed—or if it did, then only at the very beginning of her thinking.

2.2 Edith Steins Contribution to Husserlian Phenomenological Thinking and its Mirroring in her Own Works

In the foreword to the dissertation *On the Problem of Empathy*, Edith Stein explains that she has acquired a lot of ideas from Husserl's *Ideas II* and that she can no longer clearly distinguish which thoughts came from her and which from Husserl:

Since I submitted it to the faculty, I have, in my capacity as private assistant to my respected professor Husserl, had a look at the manuscript of Part II of his "Ideen", dealing in part with the same question. Thus, naturally, should I take up my theme again, I would not be able to refrain from using the new suggestions received. Of course, the statement of the problem and my method of work have grown entirely out of intellectual stimuli received from Professor Husserl so that in any case what I may claim as "my spiritual property" in the following exposition is more questionable. (Stein, 1989, p. 2)

It must be noted, however, that she refers to Husserl only once in the entire dissertation, even if her presentation shows great similarities both thematically and in terms of content with Husserl's *Ideas II*, Section II.[6] Thus, the similarity is—one could say—just formal. Husserl's influence can be seen throughout Stein's life's work in the phenomenological conceptual apparatus, but it has been said to be impossible to identify Husserl's phenomenology with Stein's phenomenological purposes, especially when it comes to Stein's later texts. This view is particularly evident in case of the Husserliana XIII, *Phenomenology of Intersubjectivity* (Husserl, 1973b), the text of which was composed by Husserl while Stein was writing her dissertation; however, according to the editorial introduction by Iso Kern, aside from the thematic agreement between the works, no influence of Stein's could be identified in Husserl's manuscripts, nor is there any evidence that Stein was familiar with these manuscripts (Husserl, 1973b, p. XLII, n. 5). Obviously, Stein had access to certain of Husserl's manuscripts during her doctoral period, while others remained completely unavailable to her. It is thus not surprising that her phenomenological knowledge provides, on the one hand, information about Husserl's current insights, while on the other hand, her dissertation demonstrates her incomplete knowledge of the manuscripts.

[6] Cf. Husserl (1952, pp. 90–169): *Zweiter Abschnitt: Die Konstitution der animalischen Natur*, which mostly corresponds to Husserl (2025, pp. 5–64): *Teilentwurf I, I. Teil.*

As was described above, the first phase of Stein's relationship with Husserl's phenomenology concluded with the ending of her assistantship. The previous 5 years (encompassing Stein's decision to pursue philosophy, her work on her dissertation, and her assistantship) show her close connection with Husserl's phenomenology. When Stein's assistantship with Husserl ended in 1918, the close ties to Husserl's phenomenology also gradually loosened. But despite of the rather ambivalent relationship between her and Husserl, she maintained contact with him and with Husserlian phenomenology between 1918 and 1922, during which time she returned to her hometown Breslau and taught for several years in a secondary school.

Stein's *Philosophy of Psychology and the Humanities*, which was published in 1922, seems to indicate a connection with the problem of Husserl's essays on "Phenomenology and Psychology" (Husserl, 1987, pp. 82–125). However, this connection is difficult to spell out in detail, apart from the general fact of Stein's assistantship. The "Contributions" (*Beiträge*, as this work is referred to in its full German title) were dedicated to Husserl on his 60th birthday and first appeared in Husserl's yearbook. In the foreword to the text, Stein indicates that her relationship to Husserlian phenomenology has changed with time and acknowledges the Husserlian influence on her work. She states that "all his [Husserl's] manuscripts from the last few years have been at my disposal" (Stein, 2000a, p. 2), but it would be difficult to identify exactly which manuscript might have influenced Stein in this work. However, it is remarkable that Stein divided her work into two parts, where the first part focuses on the relationship between psychical acts and mental constitution, while the second investigates the transition from individual acts to community acts. Since Stein hardly continues this line of thoughts in her other works, one has to suspect that Stein deliberately followed Husserlian themes at this time, with the aim of obtaining her habilitation.

Besides the work *Philosophy of Psychology and the Humanities*, Stein worked on another paper between 1917 and 1931, repeatedly revising it, which also bears some signs of Husserl's philosophy, and which has a thematical connection to *Philosophy of Psychology* too. The first phase of *Introduction to Philosophy* (Stein, 2004) goes back to her time as an assistant in 1917, which was inspired by the problem of the subject in Husserl's *Ideas II*. According to the editor of Stein's manuscript, Claudia Mariele Wulf, there are different manuscripts corresponding to the various developmental stages of the texts, which prove that the author had merged older and newer parts during preparing the final text. In the first phase (during her time as Husserl's assistant, between 1916 and 1918), Stein probably wrote a paper entitled "The ontic structure of the person and its epistemological problems" which she later used in *Introduction to Philosophy* and in *Philosophy of Psychology*. The theory of the different developmental stages is supported by a letter to Roman Ingarden from 1917, in which she mentions a work dedicated to Husserl's *Ideas*: "At the same time, I have occupied myself somewhat with my work and have written down a few supplementary thoughts that occurred to me at the time in connection with *Ideen*." (Stein, 2014a, p. 78). Another letter to Ingarden indicates that she had finished the text and planned to publish it in the 1920s: "Funny, I have also just begun to hold lectures on the introduction to philosophy on a phenomenological foundation. Not at the university but on my own *venia legendi* in my apartment." (Stein, 2014a, p. 170).

The second part of the work, which was originally entitled "The ontic structure of the person and its epistemological problems", most likely originated from the time between 1916 and 1918, when Stein was the assistant of Husserl. This paper was incorrectly identified with Stein's 1921 religious-philosophical investigation, *Freiheit und Gnade (Freedom and Grace)* (Stein, 2014b, pp. 8–73), and was first published in the *Edith Steins Werke* series. The text continues the ideas of the dissertation regarding the phenomenology of the person, but here the "pure I" has a different meaning. In the dissertation, the subject of the life of the ego has an analogous meaning to the pure ego (i.e., the pure ego was understood as the subject of experience).[7] Accordingly, the original consciousness is to be understood as exterior to the temporality of experience. Based on Husserl's argumentation about the pure I, Stein explains in *Introduction to Philosophy* that consciousness is outside of temporality, but even then, there is a paradox between the atemporality of the pure I and the source of knowledge found in the temporal experiences.[8] The pure I cannot gain knowledge from itself if the knowledge itself is a temporal category. According to Stein's understanding in the *Introduction to Philosophy*, epistemology and ontology come together at this point, "because it belongs to the essence of consciousness to be conscious, i.e., to become aware of oneself" (Stein, 2004, p. 105). This means that the ontic structure of the person as the subject of the life of the ego transcends its own temporal dimension in the act of experience. The insight of the atemporal dimension of the pure I—hence the opposition between the internal, atemporality of perception and the subject of the actual experiences of the pure I in the dissertation text (Stein, 1989, pp. 29–30)—leads Stein to the opposition between finite and eternal being in her later phenomenological writings. There, she identifies the pure I with the self in Husserlian phenomenology, which is at the same time the individual life of the subject.[9]

The third phase of Edith Stein's relationship with Husserlian phenomenology signals a return to Husserl's phenomenology after a long break, which fostered a new horizon in her thinking. From 1923 to 1931, Edith Stein worked as a

---

[7] Cf. Stein (1989, p. 29): "These experiences of my own, the pure experiences of the pure 'I', are given to me in reflection. This means that the 'I' turns back and away from its object and looks at the experience of this object." Cf. also pp. 11, 30, 39, and p. 40 where she writes: "The substantial unity is 'my' soul when the experiences in which it is apparent are 'my' experiences or acts in which my pure 'I' lives."

[8] Cf. Stein (2004, pp. 106–107), my own translation: "In original experience I have no detached 'moments', no sameness, difference or similarity with earlier experience, no 'taking' of a common essence and cognition as this or that (as perception, as joy, as resolution etc.): all this already presupposes objectification, reflective comprehension. But this can only rise on the ground of the original consciousness."

[9] Cf. Stein (2002, p. 48): "Right now the important thing is to make it clear that the pure ego is alive in every experience and cannot be eliminated from it. It is inseparable from the experiential content but nonetheless not actually a *part* of this content. Rather the converse is true: Every experience is part of the pure ego; the pure ego is alive in every experience; *its life* is that every flux in which ever new structures of experiential units arise."

teacher at the Dominican School in Speyer and elaborated on her translation of Thomas Aquinas' *Questiones disputatae de veritate*. During this time, she had little contact with Husserl's phenomenology or with Husserl personally, though, she was already working on a comparison between the phenomenological methodology and Thomas's thinking.[10] The publication of Husserl's book on time-constitution from 1928 (Husserl, 1969), about which she corresponded with Fritz Kaufmann, also gave her new impulses regarding phenomenology, and she wanted to obtain the book immediately.[11] According to Roman Ingarden,[12] Stein's return to Husserlian phenomenology is the first sign of Stein's renewed interest in phenomenology, and the studies of Husserl's work inspired a beginning or new direction in her philosophy.[13] The problems of *Ideen* have already been completely transcended by Stein here.

For the *Festschrift* which was edited for Husserl's 70th birthday in 1929 (Heidegger, 1929), Stein wrote a dialogue between Husserl and Thomas Aquinas, „Was ist Philosophie? Ein Gespräch zwischen Edmund Husserl und Thomas von Aquino", about the place of knowledge in the two philosophers' thinking (Stein, 2014c, pp. 91–119; 2000b). As is well known, Stein came up with the idea for this work through her previous translation of the work of Thomas Aquinas titled *Quaestiones Disputatae de Veritate* (Stein, 2008), which dealt with the question of being that Stein wanted to approach through phenomenology. The importance of this work becomes clear retrospectively; i.e., based on her subsequent works. In a letter to Husserl, Stein reported very vividly about her current occupations, which fascinated her after the publication of the Husserl–Thomas dialog. She wrote about her translation of Thomas, and thanked Husserl for his book *Formal and Transcendental Logic* (Stein, 2000c, Nr. 75).

Two papers originated from this period, which she wrote based on Husserl's phenomenology. In the paper, *Die weltanschauliche Bedeutung der Phänomenologie*"(Stein, 2014d, pp. 143–59), which dates from 1930/31 and illustrates the connection between world view and phenomenology, Stein discusses whether the truths of faith can also modify the scientific world-view, and how the world-view can influence philosophical thinking. In order to represent the ideological task of phenomenology, Stein referred to four works of Husserl's: *Logical Investigations, Philosophy as a Rigorous Science, Ideas for a Pure Phenomenology and Phenomenological Philosophy*, and *Cartesian Meditations*. In this paper, and in the time of the 1930s, Stein found Husserl's great merit in the problem of constitution,

---

[10] Cf. Stein's letter to Roman Ingarden in August 1, 1923, and her introductory part to her translation in *Questiones disputatae de veritate*.

[11] Cf. Stein (1993, p. 56): "Could you perhaps find out diplomatically whether Husserl has any intention of sending me a *separatum* of his article on time?"

[12] Cf. Ingarden (1999, p. 238): "The new problematic that Edith Stein is now dealing with clearly turns out to be the problematic of God, which was not mentioned before."

[13] Her renewed interest proves three contributions to the Husserlian phenomenology as *„Was ist Phänomenologie*" (1924), „Die weltanschauliche Bedeutung der Phänomenologie" 1930/31 and a review to Husserl's *Cartesian Meditations*, entitled *„Husserls transcendentale Phänomenologie*" (1931). All these contributions prove a new oriented and external approach to Husserl's thinking.

whose conclusions were still little appreciated in philosophical thinking at that time, and which provoked an utter lack of understanding when *Ideas* were published (Stein, 2014d, pp. 151–152).

The article "Husserl's transcendental phenomenology" (Stein, 2014e, pp. 159–162) was put together from two texts from 1930 and 1931. In this short article, Stein presents the advantages and developments of transcendental phenomenology, while she emphasizes the idea of an absolute and universal science in Husserl's *Cartesian Meditations*. She approaches Husserl's phenomenology epistemologically and finds that it stands in a methodological relationship with Thomistic philosophy. As she stated in the paper "Husserl's Phenomenology and the Philosophy of Thomas Aquinas", Husserlian phenomenology is determined by the method of "essence intuition" [*Wesensschau*], through the act of intention, and thus could find common ground with Thomas' sense of understanding.

The third article of Edith Stein, the genre of which is also a short review based on Husserl's later works, is entitled "Edmund Husserl. The Crisis of European Sciences and Transcendental Phenomenology" (Stein, 2014f, pp. 189–194), originates from 1937 when Edith Stein was already a Carmelite sister. At the beginning of the paper, Stein explains Husserl's working method, reflecting on her time as assistant to Husserl: "What Husserl has published was always the result of many years of research, the succinct expression of multifaceted research in which much remained unsaid that was laid down in unprinted drafts or communicated in an oral exchange of ideas." (Stein, 2014f, p. 189, my own translation). Precisely this working method of Husserl, the incompleteness of the problems, made it difficult for his students to understand his phenomenology; where Husserl found new levels of thought "which he has left behind", the "distant [observer]" only found new problems. "That is why the latest work, like many previous ones, has the character of an introduction." (Stein, 2014f, p. 189, my own translation).

As was described previously, from the 1930s onwards, Edith Stein's thinking came to the turning point. Its main topic was no longer phenomenological questions or thinking based on Husserl's phenomenology, but rather the fruitful discussion between the philosophies of different times, in the center of which she put the question of the possibility of a *philosophia perennis*. In 1930, Stein wrote to Adelgundis Jaegerschmid, her former student in Freiburg, that she "had begun" to work on Husserl's work *Formal and Transcendental Logic* and referred to a conversation with Husserl that was related to her current work.[14] The letter to Jaegerschmid probably originates from the time when Stein was planning another attempt at habilitation in Freiburg in the early 1930s. In a separate letter to Roman Ingarden, she repeatedly mentions her new work, about which she is "still not sure what will come out of it" (Stein, 2014a, p. 89).

[14] Cf. Stein (1993, p. 60): "I have had Husserl's *Formal and Transcendental Logic* since the summer vacation. I started reading it on Thursday, because of its connection to my work precisely now, and on Friday I got your letter. I had an analogous conversation with Husserl when I saw him for the first time after my conversion; I had not seen him for an interval of eight years."

The habilitation thesis entitled *Potency and Act* (Stein, 2009) from 1932 also demonstrates Stein's turn to *philosophia perennis*, while she investigates Husserl's phenomenological methodology and scholastic philosophy using the common conceptual apparatus of the two philosophical directions. For Stein, as we could have already noticed, phenomenology is removed from Husserlian thinking in its narrow sense; instead, phenomenology is here the modern philosophical means "to get knowledge about the outside world". As she explained in the foreword to the work, she felt an inner inspiration for the implementation of the dispute between Husserl's and Thomas' thinking.[15] The methodological question, which was introduced in the first chapter of *Potency and Act* through the common conceptual apparatus, was deepened by the analyses of the formal- and material-ontological meaning of act and potency in relation to being. Stein obviously uses the material and formal ontologies in the sense of Husserl's *Ideas I* (Husserl, 1976, pp. 10–39), and she tries to prove here that the scholastic empty forms, the transcendentals, which reproduce beings in the greatest generality, also occur as the basic forms in Husserl's formal ontology. According to Thomas, the empty forms have a general content that is filled out with the concrete individual. In Stein's argumentation, thus, the phenomenological empty form is not the most general form, but the one specified and classified by an individual:

In actu esse and in *potentia esse* are modes of being, material fulfillments of the form of being, hence as such they need to be elucidated in a material ontology. But at the same time they designate formal differences of being; namely, "*in potentia esse*" designates a being that refers to another (actual) being. These modes of being are to be elucidated formally as well as materially only in connection with the being they correspond to. We have in fact seen that distinction of being run parallel to distinctions of objects and of what these objects are. We must suppose from the start that not just any object can be actual or potential (Stein, 2009, pp. 51–52, translation slightly adapted).

Stein refers here to Husserl's fundamental distinction between "generalization and formalization" in §13 of *Ideas I*, with regard to the formal and material categories. The formal categories include all "empty forms" coming from formal ontology, while the material categories have the genera of the being that define for us the branch of disciplines about being. If every being has its own measure of being, then the distinction of the categories must at the same time be distinctions of being (Stein, 2009, p. 92, translation slightly adapted). Related to the transcendentals in Thomistic philosophy, Stein distinguishes transcendentals from the categories, "in which the transcendentals do not fit; they are: *aliquid*, *ens* [being], *unum*, *verum*, and *bonum*. We can understand the being in such a way that it does not enter the table of categories, but is presupposed by it; that is, in the general sense wherein both the self-sufficient and the un-self-sufficient 'are'. The being thus comes before the division

[15] Stein (2009, p. 3): "The author, whose philosophical thinking was formed by Edmund Husserl, has in recent years found a home in Aquinas's thought world. She now feels an inner need to allow the two different ways of doing philosophy bearing their names to come to resolution within herself."

of substance and accident." (Stein, 2009, p. 91). This is the place where Stein first formulates *philosophia perennis*, the question of being, which precedes every other philosophical question and thus provides the ground for both phenomenological and scholastic thinking.

The habilitation thesis thus indicates a new direction in Stein's thinking. On the one hand, she attempts to define and to unite the two schools of thought through parallel conceptual analysis, while on the other hand, she transcends the border of classical phenomenological thinking by extending her methodology to the question of being in the *Excursus* of *Potency and Act* about Husserl's transcendental idealism. Although even in her last philosophical work, *Finite and Eternal Being*, Stein characterizes her philosophical direction as being affected by Husserl's phenomenology, her relationship to phenomenology is limited to generalities, such as that phenomenology is still to be considered as the modern philosophical trend, deriving its name from Husserl's school of thought (Stein, 2002, XXVIIff). This gradual withdrawal can be observed in Stein's scientific relationship with Husserl too, while their personal relationship does not show the same changes over time. In Stein's later work, the general apparatus of terms suggests an independent yet engaged relationship with Husserl's phenomenology. It cannot be equated with Husserl's own phenomenological aims; rather, it reveals her individual religious phenomenological approach, which later resonates more strongly in her phenomenological thinking.

References

Heidegger, M. (1929). JPPF. E. Husserl zum *70. Geburtstag gewidmet.* Spec. Issue 10.

Husserl, E. (1928). Vorlesungen zur Phänomenologie des inneren Zeitbewusstseins. In M. Heidegger (Ed.), *JPPF* (Vol. 9, pp. 369–450). Max Niemeyer.

Husserl, E. (1952). *Ideen zu einer reinen Phänomenologie und phänomenologischen Philosophie. Zweites Buch: Phänomenologische Untersuchungen zur Konstitution, Hua IV* (M. Biemel, Ed.). Martinus Nijhoff.

Husserl, E. (1969). *Zur Phänomenologie des inneren Zeitbewustseins, Hua X* (R. Boehm, Ed.). Kluwer.

Husserl, E. (1971). *Ideen zu einer reinen Phänomenologie und phänomenologischen Philosophie. Drittes Buch: Die Phänomenologie und die Fundamente der Wissenschaften, Hua V* (M. Biemel, Ed.). Martinus Nijhoff.

Husserl, E. (1973a). *Ding und Raum. Vorlesungen 1907, Hua X* (Ulrich Claesges, Ed.). Kluwer.

Husserl, E. (1973b). *Zur Phänomenologie der Intersubjektivität. Texte aus dem Nachlass. Erster Teil: 1905–1920, Hua XIII* (I. Kern, Ed.). Kluwer.

Husserl, E. (1976). *Ideen zu einer reinen Phänomenologie und phänomenologischen Philosophie. Erstes Buch: Allgemeine Einführung in die reine Phänomenologie, Hua III* (K. Schumann, Ed.). Martinus Nijhoff.

Husserl, E. (1987). *Aufsätze und Vorträge (1911– 1921), Hua XXV* (T. Nenon & H. R. Sepp, Eds.). Kluwer.

Husserl, E. (1994). *Wissenschaftskorrespondenz. HuaDok III* (Vol. 7, K. Schumann & E. Schumann, Eds.). Kluwer.

Husserl, E. (2001). *Die Bernauer Manuskripte über das Zeitbewusstsein (1917/18) Hua XXXIII* (R. Bernet & D. Lohmar, Eds.). Kluwer.

Husserl, E. (2002). *Urteilstheorie. Vorlesung 1905 HuaMat V* (E. Schumann, Ed.). Kluwer.

Husserl, E. (2025). *Ideen zu einer reinen Phänomenologie und phänomenologischen Philosophie. Zweites Buch: Phänomenologische Untersuchungen zur Konstitution und Wissenschaftstheorie, Hua IV/V* (Dirk Fonfara, Ed.). Springer.
Ingarden, R. (1999). Über die philosophischen Forschungen Edith Steins. In W. Galewicz (Ed.), *Ingarden, Roman. 1999. Schriften zur frühen Phänomenologie, Gesammelte Werke 3* (pp. 228–255). Max Niemeyer.
Jani, A. (2015). *Edith Steins Denkweg von der Phänomenologie zur Seinsphilosophie.* Königshausen & Neumann.
Schumann, K. (1977). *Husserl-Chronik. Denk- und Lebensweg Edmund Husserls.* Martinus Nijhoff.
Schumann, K. (1991). Einleitung zu Husserls Exzerpt aus der Staatsexamenarbeit von Edith Stein. *Tijdschrift voor Filosofie, 53*, 687–699.
Stein, E. (1989). *On the problem of empathy* (W. Stein, Trans.). ICS Publications.
Stein, E. (1993). *Self portrait in letters 1916–1942, CWES V* (J. Koeppel, Trans.). ICS Publications.
Stein, E. (1998). Exkurs über den transzendentalen Idealismus. In H. R. Sepp (Ed.), *Potenz und Akt. Studien zur Philosophie des Seins. ESGA 10* (235–236). Herder.
Stein, E. (2000a). *Philosophy of psychology and the humanities CWES VII* (M. C. Baseheart & M. Sawicki, Trans.). ICS Publications.
Stein, E. (2000b). *Knowledge and faith. CWES VIII* (W. Redmond, Trans.). ICS Publications.
Stein, E. (2000c). *Selbstblidnis in Briefen I. ESGA 2* (M. A. Neyer, Ed.). Herder.
Stein, E. (2002). *Finite and eternal being. An attempt at an ascent to the meaning of being. CWES IX* (K. F. Reinhardt, Trans.). ICS Publications.
Stein, E. (2004). *Einführung in die Philosophie, ESGA 8* (C. M. Wulf, Ed.). Herder.
Stein, E. (2008). *Übersetzung: Des Hl. Thomas von Aquino Untersuchungen über die Wahrheit – Questiones disputatae de veritate 1–2, ESGA 23/24* (A. Speer & F. Valerio Tommasi, Eds.). Herder.
Stein, E. (2009). *Potency and act, CWES XI* (W. Redmond, Trans.). ICS Publications.
Stein, E. (2014a). *Letters to Roman ingarden, CWES XII* (H. Candler Hunt, Trans.). ICS Publications.
Stein, E. (2014b). Freiheit und Gnade. In B. Beckmann-Zöller & H. R. Sepp (Eds.), *Freiheit und Gnade und weitere Beiträge zu Phänomenologie und Ontologie, ESGA 9* (8–73). Herder.
Stein, E. (2014c). Was ist Philosophie? Ein Gespräch zwischen Edmund Husserl und Thomas von Aquino. In B. Beckmann-Zöller & H. R. Sepp (Eds.), *Freiheit und Gnade und weitere Beiträge zu Phänomenologie und Ontologie, ESGA 9* (91–119). Herder.
Stein, E. (2014d). Die weltanschauliche Bedeutung der Phänomenologie. In B. Beckmann-Zöller & H. R. Sepp (Eds.), *Freiheit und Gnade und weitere Beiträge zur Phänomenologie und Ontologie, ESGA 9* (143–159). Herder.
Stein, E. (2014e). Husserls transzendentale Phänomenologie. In B. Beckmann-Zöller & H. R. Sepp (Eds.), *Freiheit und Gnade und weitere Beiträge zur Phänomenologie und Ontologie ESGA 9* (159–162). Herder.
Stein, E. (2014f). Edmund Husserl. Der Krisis der europäischen Wissenschaften und die transzendentale Phänomenologie. In B. Beckmann-Zöller & H. R. Sepp (Eds.), *Freiheit und Gnade und weitere Beiträge zur Phänomenologie und Ontologie, ESGA 9* (189–194). Herder.
Stein, E. 2016. *Life in a Jewish family CWES I* (J. Koeppel, Trans.). ICS Publications.

Anna Jani has been a senior researcher at the Makovecz Campus Foundation in Piliscsaba, Hungary, since 2022. Her main research interests include early phenomenolgy, the phenomenology of religion, and hermeneutics. Selected monographs: *Ways in the Phenomenology of Religion: Adolf Reinach, Max Scheler, Martin Heidegger, Edith Stein* (in Hungarian), MCA – Collegium Professorum Hungarorum 2024; *The Ontological Roots of Phenomenology. Rethinking the History of Phenomenology and Its Religious Turn*, Lexington Books 2022; *Edith Steins Denkweg von der Phänomenologie zur Seinsphilosophie*, Königshausen & Neumann 2015; Private homepage: www.annajani.info.

Chapter 3
Husserl and Stein on Empathy: Paving a Way Towards Integration

Francis B. Payo

Abstract The interconnection of Husserl and Stein, in view of their lives and selected texts (Stein's *On the Problem of Empathy* and Husserl's *Ideas II* and *Cartesian Meditations*), shows how they influenced each other and how they worked collaboratively in advancing the phenomenological project. Such a relationship supports the case for comparison and integration of their phenomenological accounts of empathy. Given such a context, we ask: *How can we delineate the areas where they converge and diverge in constituting empathy? Can we find a way to integrate their phenomenological understanding of empathy?*

This paper presents some findings of a lengthy process of work in response to such questions. First, Husserl and Stein each have their unique contribution to the phenomenological understanding of empathy, and they have influenced each other's understanding along the way. Second, their different ways of understanding empathy, inevitably affecting the constitution of intersubjectivity, led Husserl and Stein in different directions. Third, Husserl and Stein converge more than diverge in their understanding of empathy as embodied in their specific texts. These salient aspects pave a way for integration, and perhaps a richer understanding of empathy. I find such integration prominently explicit and feasible at the basic phenomenological understanding of empathy.

Keywords Empathy · Phenomenology · Intersubjectivity · Edith Stein · Edmund Husserl · Convergence/Divergence · Integration/Synthesis

3.1 Introduction

Since I do not have the luxury of space to include the lengthy account of such study (Payo, 2013), this brief account presents only an overview of some salient aspects. Here, I introduce Edmund Husserl and Edith Stein by highlighting some significant

F. B. Payo (✉)
Manila, Philippines
e-mail: francis.payo.eccle@ust.edu.ph

M. Lebech (ed.), *Stein and Husserl's Phenomenology of the Social in Ideas II*, Women in the History of Philosophy and Sciences 25,
https://doi.org/10.1007/978-3-032-17339-3_3

aspects of the life, personality and thought of each of them, as these have a large bearing on the course of my investigation. I also introduce their works in general, but focusing on specific works: namely, Stein's *On the Problem of Empathy* (1989) and Husserl's *Cartesian Meditations* (1977) and *Ideas II* (1989). These considerations provide the general and specific context necessary to understand their respective accounts of empathy. With such historical and theoretical context, I carried out the articulation of their respective understandings of empathy, first of Stein's and then of Husserl's. My consideration of these two understandings of empathy from the phenomenological standpoint further ushered me to the next crucial step: that of pointing out areas where they converge and diverge. I have only indicated areas which are evident in their clarity. Such indications led me to discern a path towards a possible integration of the two accounts of empathy. I have not thoroughly explored such a path, however; only mapped it out, making signposts, thus paving the way towards integration. A concrete attempt to integrate it thoroughly is a further step yet to be taken.

3.2 Considering Husserl's and Stein's Lives and Texts

Husserl had already mentioned "empathy" long before Stein got hold of it.[1] But it was Stein, inspired and directed by Husserl, who first developed the notion of "empathy" via phenomenological constitution in her dissertation (Lebech, 2008).[2] Stein expresses her indebtedness to Husserl by acknowledging that her problem and method "have grown entirely out of intellectual stimuli received from Professor Husserl."[3] Husserl's manuscripts of *Ideas II*, written before Stein's dissertation, prove that Husserl had already an account of empathy.[4] Husserl and Stein together wrestled with the notion of "empathy" in *Ideas II* and other works, when Stein served as assistant to Husserl (Sawicki, 1997).[5] In general, they share the same basic understanding

[1] See Edmund Husserl, *Ideas I*, 75, par. 2. Cf. MacIntyre, 67.

[2] By "via phenomenological constitution," I refer in general to how Stein applies the phenomenological method in analyzing empathy. Mette Lebech, in particular, succinctly describes Stein's dissertation as "an eidetic analysis of the essence of empathy and two constitutional analyses pertaining to what empathy contributes to the constitution of the pyscho-physical individual." Cf. M. Lebech, "Stein's Phenomenology of the Body: The Constitution of the Human Being Between Description of Experience and Social Construction," *Maynooth Philosophical Papers*, Issue 5 (2008), ed. Simon Nolan, 16. Berkman (2006) points out that Stein's interest in empathy evolved into her dissertation topic and owed much to her study of Husserl's philosophy of intersubjectivity; 28.

[3] Stein, *On the Problem of Empathy*, 1.

[4] Cf. Thomas Nenon, "Husserl's Theory of the Mental," in *Issues in Husserl's "Ideas II,"* ed. Thomas Nenon (Netherlands: Kluwer Academic Publishers, 1996), 223ff.

[5] Stein was thereby exposed to Husserl's *Ideas II* which contains a thorough account of empathy, considering that such manuscripts precede Stein's dissertation. This shows that Husserl had already worked out his own account of empathy though not available yet (since it was published many years later) for Stein's consideration. Cf. Payo, 104.

of empathy in their early works on constitution (Calcagno, 2007).[6] Husserl made his definitive account of intersubjectivity (which necessarily includes "empathy") in *Cartesian Meditations* in view of an epistemological concern of establishing the foundations of sciences, both in the realm of nature and in the realm of spirit.

Husserl's and Stein's mutual influence on each other manifests in their work of collaboration in the continuing development of phenomenology. Her work shows that Stein was not a mere student or assistant to Husserl, but a collaborator. Their intermingled lives and texts testify to such work of collaboration between Husserl and Stein, although the latter's contribution has been neglected until quite recently.

Thus, the interconnection of Husserl and Stein, in view of their lives and texts, shows how they influenced each other and how they worked collaboratively in advancing the phenomenological project. Their relationship of mutual influence and collaboration helps make the case for a comparison and integration of their phenomenological accounts of empathy.

3.3 Considering Husserl's and Stein's Convergent and Divergent Insights

Husserl's and Stein's respective phenomenological accounts of empathy[7] exemplify their influence on one another and their collaboration, although the extent of such influence and collaboration needs to be explored even more, if only to enrich ourselves with their insights, both where they converge and where they diverge. I find more areas of convergence, in the broad outlines and trajectory of phenomenology, as well as in the details of their constitutional analyses of empathy. Such findings show not only Stein's fidelity to Husserl's phenomenological project in general, but also her independence of thought in her specific direction and constitutional analyses.

3.3.1 In the Application of the Phenomenological Method

Both Stein and Husserl applied the phenomenological method *rigorously* in their descriptions and analyses of phenomena (e.g., empathy); i.e., constituting our "lived world" to yield us a legitimate and truthful account of reality. While Husserl applied the phenomenological method quite exclusively, Stein applied it along with the

---

[6] Antonio Calcagno, after presenting the structural analysis of Stein's *Introduction to Philosophy* and Husserl's *Ideas II* and *Ideas III* remarks how such convergences "reveal an intimate relationship between the ideas of the two thinkers." He follows this up with a discussion of how "Stein follows Husserl very closely on certain key insights." Among those themes which bear striking similarities for both thinkers, Calcagno mentions their analyses of empathy.

[7] I considered Husserl's and Stein's accounts of empathy in chapters 2–3 of a lengthy work already mentioned above, a dissertation titled *Edith Stein and Edmund Husserl on Empathy: Towards a Comparative Synthesis* (2013), 112–299.

approaches and insights she got from Dilthey, Scheler, and also St. Thomas Aquinas later in her life.[8] However, in spite of the fact that they both used the same phenomenological method, they could vary in their actual constitutional analyses using their respective intuition (Sweet & Feist, 2003, 4–5).[9]

While both stayed within the ambit of phenomenology, Husserl's ultimate goal is transcendental phenomenology; that is, he has a predominantly epistemological focus (for the foundation of sciences),[10] which is shared by Stein in general. However, Stein also pursues the understanding of the human person and community (dwelling more deeply on religious and spiritual dimensions),[11] thereby taking a predominantly personalist approach. Husserl's constitutional analyses also include clarifications of the sense of community; thus, these concerns regarding persons and communities (close to Stein's heart) are also at the heart of Husserl's epistemological considerations. Such application of the phenomenological method has a significant bearing on their respective accounts of empathy.

[8] The influences of Scheler and Dilthey on Stein are unmistakable in her dissertation. Stein confidently agrees and disagrees with Scheler and Dilthey, but most of all, she confidently formulates or articulates her own position—neither completely opposed nor completely in agreement with them. This seems true as well with Husserl.

[9] Intuition is essential in the phenomenological approach, hence being considered as the "life-blood" of Husserl's approach. Husserl and Stein consistently made use of intuition in their constitutional analyses, and they applied it as if it was the only kind of reasoning available to them. Once they begin using the phenomenological method, all the results of deduction are to be bracketed by default, since such results stand on certain presuppositions. If phenomenology exemplifies a *presuppositionless* science, then what is left to start with, after everything has been bracketed, is *the Ego, which always has the capacity for intuition.* Hence, intuition is the only available means of reasoning that allows Husserl and Stein to proceed with their phenomenological clarification, first of oneself, and then of the surrounding world. Therefore, on the basis of their capacity for intuition, they may vary in their insights on the matter under consideration, which forms the basis for some divergence.

[10] Sawicki notes that Husserl was "working on the construction of scientific rationality itself," obviously epistemologically directed. Cf. *Body, Text, and Science: The Literacy of Investigative Practices and the Phenomenology of Edith Stein* (Dordrecht/Boston/London: Kluwer Academic Publishers, 1997), 103. In a similar vein, Angela Ales Bello says that Husserl "felt compelled to perfect his method as an essentially cognitive instrument." See "Ontology, Metaphysics, and Life in Edith Stein," trans. Antonio Calcagno, *Contemplating Edith Stein*, ed. Joyce Avrech Berkman (Notre Dame, Indiana: Notre Dame University Press, 2006), 272.

[11] Stein's constitutional analyses on community as a reality and its fundamental relationship to the individual appear largely in her treatise on "Individual and Community," *Philosophy of Psychology and Humanities*, trans. Mary Catharine Baseheart and Marianne Sawicki, ed. Marianne Sawicki (Washington, D.C.: ICS Publications Edition, 2000). She clarifies that "communities" belong to our lived experiences, confronting us as realities in our environment. She clarifies: "A sense-bound world unfolds for the community within its experiencing. Again, it's the individuals whose mental doing is constitutive for the world of the community; but then again, not everything that belongs to their individual world gets into the community's world as well." After a sustained analysis, she concludes her clarifications here with this insight: "Thus, all social life and all social modes finally refer back to the core of the person, which is beyond the reach of all the influences of reciprocal communication."

3.3.2 *In Terms of Their Accounts of Empathy*

There are areas of *convergence and divergence* in Husserl's and Stein's understanding of empathy. However, I argue that the two accounts converge more than they diverge.

3.3.3 *Empathy as Description and Possibility of the Other*

Both Husserl and Stein offer a phenomenological description of empathy which includes a clarification of the existence-status of the other. Husserl describes empathy as the experience of someone else: "Thus, the problem is stated at first as a special one, namely that of the 'thereness-for-me' of others, and accordingly as the theme of a transcendental theory of experiencing someone else, a transcendental theory of so-called 'empathy'."[12] Stein describes empathy as the perceiving of foreign subjects and their experience: "I recognized this basic problem to be the question of empathy as the perceiving [*Erfahrung*] of foreign subjects and their experiences [*Erleben*]."[13] For Stein, empathy is an act "in which foreign experience is comprehended."[14] Thus, both Husserl and Stein take into account a description of empathy and their application to psycho-physical-spiritual phenomena, which takes account necessarily of the possibility of the other in its essence and existence. Such is the givenness of our lived experiences pertaining to an other. The work of constitution clarifies precisely how such lived experiences arise or appear in our consciousness. A point of divergence is raised by Sawicki regarding Stein's empathy as a rival alternative to Husserl's constitution (Sawicki, 1997).[15]

---

[12] Edmund Husserl, *Cartesian Meditations: An Introduction to Phenomenology*, trans. Dorion Cairns (The Hague: Martinus Nijhoff, 1977), 43.

[13] Edith Stein, *On the Problem of Empathy*, 3rd revised edition, trans. Waltraut Stein, from *The Collected Works of Edith Stein*, vol. 3 (Washington, D.C.: ICS Publications, 1989), 1.

[14] Stein, *On the Problem of Empathy*, 6.

[15] Sawicki raises an interesting point by distinguishing constitution from empathy, the way Stein proposes it (though reluctantly) as a rival alternative to constitution: "the free I as live body" can be constituted, but not the "creativity of deciding acts," which can be grasped only through empathy. Thus, the "embodied I can be constituted as an individual: it is some body or other, no matter who. But the personal creative I can be known for who she is only with empathy." I find this worthy of further verification. Sawicki further distinguishes Husserl from Stein: "Unlike Husserl, you have shown that it is empathy that gives rise to the plurality of centers, and therefore is the foundation for the real world and for individual identity alike. Empathy is logically prior to Husserlian constitution." See 123, 127.

3.3.4 Basic Description or Understanding of Empathy

In the most general sense, Husserl describes empathy as the experience of someone else, while Stein describes empathy as the perceiving of foreign subjects and their experiences. In short, it is basically one's experience of another; i.e., my experience of others psycho-physically and spiritually. Crucial to such understanding is the distinction between primordiality (also called originary or primal presence) and non-primordiality (non-originary presence, co-presence or appresence). Empathy, according to Stein, is "an act which is primordial as present experience though non-primordial in content."[16] For Husserl, "it is characteristic of empathy that it refers to an originary Body-spirit-consciousness but one I cannot myself accomplish originarily, I who am not the other and who only function, in regard to him, as a comprehending analogon."[17] Both Stein and Husserl understand an act of empathy as the combination of primordial and non-primordial elements, although they may have used different terminology. Husserl prefers to use the terms "association", "mirroring" or "pairing", but also uses the term "combination". Husserl uses "appresentation" to describe an act of empathy as making the other present to me—i.e., a non-original presence is combined with my original presence.[18] The primordial and non-primordial character of the perception which is characteristic of empathy is what sets it apart from other acts of perception; hence, for Stein, empathy is a kind of act of perceiving *sui generis*.[19] Thus, it is clear that both Husserl and Stein share a basic understanding of empathy, considering their different ways of clarifying the same lived experience (Zahavi, 2010).[20] We find how they converge in understanding how empathy operates at the bodily level as well as at the psychic and spiritual realms,

---

[16] Stein, *On the Problem of Empathy*, 10. Sawicki clarifies this: "Obviously, the reflective act of having an empathy is indeed an originary experience in its own right and I feel myself active in it. The non-originary experience appears as *the intentional content* of that originary act, and the non-originarity of the content consists in its having registered with me as something in which another I first lived. It is this registration that you mean to designate with the technical term *Einfuhlung.*" 101.

[17] Husserl, *Ideas II*, sec. 51, p. 208. In other words, it is impossible for me as ego to directly access an alter ego, but it is possible for me to access the alter ego indirectly and non-originally. This necessarily involves a mediating intentionality, allowing the alter ego to be present to me. The act of making the alter ego present to me is called appresentation. See also a discussion by David Carr, *Interpreting Husserl: Critical and Comparative Studies* (Dordrecht/Boston/Lancaster: Martinus Nijhoff Publishers, 1987), 54ff., and A.D. Smith, *Husserl and the Cartesian Meditations* (London and New York: Routledge, Taylor & Francis Group, 2003), 228–229.

[18] Husserl, *Cartesian Meditations*, sec. 50, 109; *Ideas II*, sec. 44 and 51, 171–208.

[19] Stein, *On the Problem of Empathy*. Stein clarifies the phenomenon of empathy in this way: "And while I am living in the other's joy, I do not feel primordial joy […] This other subject is primordial although I do not experience it as primordial. In my non-primordial experience I feel, as it were, led by a primordial one not experienced by me but still there, manifesting itself in my non-primordial experience." 11.

[20] According to Dan Zahavi, they both share the understanding that "empathy is a basic, irreducible form of intentionality that is directed towards the experiences of others." He claims that not only Husserl and Stein share this understanding, but Scheler as well. Cf. "Empathy, Embodiment and Interpersonal Understanding: From Lipps to Schutz," *Inquiry*, vol. 53, no. 3 (June 2010), 291.

which is inevitable because the person is a physico-psycho-spiritual unity (Calcagno, 2007).[21]

3.3.5 The Importance of the Body in an Act of Empathy

For both Husserl and Stein, the body is crucial in the constitution of the individual subject or person.[22] In the process of reduction, Husserl points out an important discovery: "This abstractive sense-exclusion of that which is alien leaves us a *kind of 'world'* still, a Nature reduced to what is included in our ownness and, as having its place in this Nature, thanks to the bodily organism, the psychophysical Ego, with 'body and soul' and personal Ego – utterly *unique* members of this reduced 'world'."[23] Husserl makes us see that in perceiving a man, for example, we first of all have the material corporeal body—the lowest, most basic level in experiencing the other, but also foundational and therefore, crucial and indispensable. Psychic existence, for Husserl, is appresented in corporeality, and thus he concludes: "Therefore, under the title 'other man', we have a Body, and we have this Body in union with sensorial fields and, as it were, psychic fields, i.e., in union with a subject of acts."[24] In other words, our fundamental recognition of another subject is as a body. For her part, Stein posits as indubitable (in the process of reduction) the phenomenon of "a world of physical bodies and experiencing subjects external to us," as well as the phenomenon of "foreign psychic life."[25] What is characterized as primordial is so precisely because of its embodied givenness, which is spatio-temporal in character. In other words, Stein characterizes primordiality as bodily presence, either in terms of being given primordially or primordial in content.[26] The body, as that which provides the primordial character of the perceiving experience, seems to be taken here as the most basic level in experiencing the other as someone else or as a foreign individual, and is thus apparently indispensable. Mette Lebech, however, offers a plausible insight contrary to considering the body as the most basic level

---

[21] Calcagno puts it this way: "The person is seen as a bodily psycho-spiritual unity, and hence what is perceived on a bodily level is related to the psycho-spiritual level. Just as there is a transferred co-presence of my body and the body of the other, so too there is a transferred co-presence of my ego with another's ego." *The Philosophy of Edith Stein* (USA: Duquesne University Press, 2007), 140-141. See Husserl, *Ideas II*; Husserl, *Cartesian Meditations, Fifth Meditation*; Stein, *On the Problem of Empathy*.

[22] Elizabeth Behnke asserts the importance of the body even prior to constitution. See "Phenomenology of the Body in *Ideas II*," *Issues in Husserl's Ideas II*, ed. Thomas Nenon and Lester Embree (Netherlands: Kluwer Academic Publishers, 1996), 135ff., 156.

[23] Husserl, *Cartesian Meditations,* sec. 44, p. 98.

[24] Husserl, *Ideas II*, sec. 45–46, p. 17.

[25] Stein, *On the Problem of Empathy*, 5.

[26] Stein, *On the Problem of Empathy*, 7.

of experiencing the other and thus taking it as indispensable.[27] In the domain of psycho-physical individuals, the body serves as the locus of the I's sensations and psychic acts, which are then perceived as the basis for perceiving an other in an act of empathy. Thus, empathy brings to light the indispensable importance of the body in our continuous acts of perception, in our personal encounters. Stein's divergence from Husserl, according to Sawicki, lies in this: While following Husserl's description of the body as "sentient space-organizer", Stein emphasizes irreplaceability while Husserl emphasizes "literal *interchangeability* of bodies" (Sawicki, 1997). I intend to mention this only, but I believe it demands clarification.

3.3.6 The Givenness of the Other to the I

It is clear in both Husserl's and Stein's texts how the issue of *givenness* is crucial in their constitutional analyses. They both criticize and offer an alternative to Theodor Lipps' account of empathy, which implied a oneness between my I and the I of the self physically separated from me. In so doing, both Husserl and Stein maintain the distance between the I and the other which can be lost in such complete union. They both constitute the I and the other as becoming identified, but only to a certain extent according to its givenness. Thus, for Husserl and Stein, the object can be experienced "more or less directly"; that is, "it can be more or less present."[28]

Both thinkers carry out the work of constitution, which is mainly a work of clarifying what is *given* in our consciousness, thereby examining how "things" arise in our consciousness. It is precisely in clarifying the intentionalities and motivations that something appears in our consciousness. While Husserl mainly uses the term "intentionality", Stein uses "motivation"—referring to the same thing. The issue of givenness of the Other to the I is crucial to Husserl's and Stein's understandings of empathy. They share the same insight that the other as a subject of sensuous

[27] M. Lebech clarifies: "If Stein thought the body was identified before its identification by constitution, why would she make a constitutional analysis of it in Chapter 3 of *On the Problem of Empathy*? 'You' is simply experienced as other as soon as the I experienced is not mine. It does not for that require a body. Husserl, however, thinks otherwise. That is the reason why Stein does not think of the body as the most basic level in experiencing the other. The very last section of *On the Problem of Empathy* raises the question as to whether the body is indispensable for empathy. She is answering: 'it is not clear' in the very last sentence. Thus she cannot be said to claim the body to be indispensable." This is from my direct correspondence with Lebech in her editorial comment of this manuscript. In response to this, let me point out that in this last section referred to by Lebech, Stein concludes her consideration by clarifying a crucial point: that we, as psycho-physical individuals, know about the spiritual life of other individuals only through the medium of corporeality, as we are always bound to a physical body. As regards purely spiritual persons, however, Stein points to "the essential possibility of genuine experience" in such realm. It is precisely this domain (not the domain of psycho-physical individuals) that Stein prefers not to enter at the moment and is content to conclude her account of empathy with the declaration of *non liquet. On the Problem of Empathy*, 38.

[28] Cf. Zahavi, "Empathy, Embodiment and Interpersonal Understanding," 293.

body is directly given to the I as a subject of sensuous body.[29] That is to say, the I's body directly experiences the other's body, through sensations. But the other as a subject of psychic reality (and spiritual as well) is not directly given to the I, only indirectly through empathy. In other words, I directly experience (through sensations) the other's body, but indirectly experience his interiority or inner life through empathy. I see Husserl and Stein as converging clearly on this.

There are various and conflicting points of view to be considered in trying to draw out precisely a discernible divergence of Stein from Husserl regarding the extent of the direct and indirect givenness in empathic acts (Hutt, 2009; Sawicki, 1997). For example, a contrary opinion by Hutt contends that they both agree that the other is given to the I for perception, but the extent of givenness is what distinguishes them from each other. Curtis Hutt insists that for Husserl, no direct awareness of the other is possible, while for Stein, it is possible and made possible by empathy. Stein's explication of empathy reveals how she wrestled with it and maintained her position accordingly. Thus, while Husserl maintains that the I cannot experience the other directly, Stein maintains that the I actually can experience the other directly.[30] I differ in opinion from Hutt on this. Husserl, having constituted the original and appresented spheres, sees an abyss between them, yet he finds that actually the sensuously seen body is experienced "forthwith" as the body of someone else, not merely an indication of someone else. Husserl sees that as he penetrates into the being of someone else, the latter "experiences me forthwith as an Other for him, just as I experience him as *my* Other."[31] For me, that shows an important point about the issue of givenness in Husserl: that he sees direct givenness of the sensed body to the sensing subject, here implied by the term "forthwith". Thus, Husserl allows such direct givenness to an extent. In the realm of psychic reality, however, Husserl sees the givenness as indirect and mediated via empathy.[32]

Furthermore, according to Hutt, Stein maintains that individuals are still radically open, prior to any rational mediation, to the emotions and thoughts of others. They come ready-made for the intersubjective and for life in communities. While Stein does not assert that immediate knowledge of the experiences of another's actual and present self are attainable, through empathy one is able to identify in a very strong sense with the thoughts and feelings of others.[33]

Sawicki also argues how Stein diverges from Husserl on this: "Constitution, as Husserl displays it, yields objects in their essential unity; but empathy yields the

[29] Cf. Husserl, *Cartesian Meditations*, sec. 42ff. This is the realm of what Husserl calls "proto-empathy", where my body and the body of the other are mutually and directly experienced, which brings to the surface at this primal level the intersubjective character of our experience, of our act of empathy.

[30] Curtis Hutt argues for this position thoroughly in his essay, "Identity, Alterity, and Ethics in the Work of Husserl and His Religious Students: Stein and Levinas," *Philosophy Today*, vol. 53, no. 1 (Spring 2009), 14–17.

[31] Husserl, *Cartesian Meditations*, sec. 56, p. 130.

[32] Husserl, *Cartesian Meditations*, sec. 55ff.

[33] Cf. Hutt, 16–17.

non-objectifiable essential character of a person appearing in 'fingerprints' distinguishing her actions and experiences [...] Your covert accomplishment is to explicate this egoic function, empathy (*Einfuhlung*), on the model of Husserlian 'constitution' but as an alternative principle of essence discernment. The essences that it discerns will be individual human characters, and it will permit their creativity and responsibility to be experienced from within, albeit nonoriginarily."[34] For Sawicki, this is Stein's radical departure from Husserl, an insight for further clarification. In addition, according to Hutt (and contrary to Stein's position regarding primordial identity), Husserl critically maintains "the separation of the transcendental subject from what is other, in spite of the early move to empathize." In so doing, "the absolute alterity (otherness) of the other is affirmed while at the same time an inevitable 'mine-ness' obtains".[35] This, too, needs clarification.

3.3.7 Nature and Spirit

The realm of nature and the realm of spirit share the same world, yet seen or constituted differently. In Stein's constitution of the psycho-physical individual, consciousness appears "not as a causally conditioned occurrence," but also as "object-constituting" – stepping out "of the order of nature" facing it. Thus, "consciousness as correlate of the object world is not nature, but spirit."[36] Nature corresponds to natural sciences, while spirit corresponds to human sciences. This most general division or distinction of reality, as seen phenomenologically, is shared by both Husserl and Stein. Causality is identified as the form of rational connectedness that governs the realm of nature, while motivation is identified as the form of rational connectedness that governs the realm of spirit. Hernandez (2008) describes the former (nature) as "the understanding of natural causality" while the latter (spirit) as "the grasping of motivations." In causality and motivation, we find two kinds of rational lawfulness: in the realm of nature and in the realm of spirit or world of values, respectively. Ullrich Melle (1996) tells us that "the relation between nature and spirit belongs to the central topics of Husserl's phenomenological philosophy."[37] Lebech points out a divergence of Stein from Husserl, citing the Final Consideration ("*Schlussbetrachtung*") of *Philosophy of Psychology and the Humanities*: "Stein does not, unlike Husserl, distinguish between natural and personalistic attitudes. To her nature is identified from the fact that the lawfulness governing it is causality and spirit, in contrast, is identified by the fact that the lawfulness governing it is motivation."[38]

[34] Sawicki, 113.

[35] Hutt, 17.

[36] Stein, *On the Problem of Empathy*, 91.

[37] Ullrich Melle, "Nature and Spirit," *Issues in Husserl's Ideas II,* ed. Thomas Nenon (Netherlands: Kluwer Academic Publishers, 1996, 16.

[38] This forms part of the editorial comment by Lebech of this manuscript, which I have integrated in the revised version.

Another divergence which Baseheart (1991) highlights is Stein's "strikingly original" contribution in constituting the spiritual person (and thereby the realm of the Spirit): "in that she arrives at rationality by way of analyses of feelings."[39] Indeed, for Stein, "consciousness as correlate of the object world is not nature, but spirit;" and in this sense, "all outer perception is carried out in spiritual acts. Similarly, in every literal act of empathy, i.e., in every comprehension of an act of feeling, we have already penetrated into the realm of the spirit," the world of values, as the new object realm constituted in feeling.[40] Therefore, as physical nature is constituted in perceptual acts, so is the spiritual world constituted in feeling. Baseheart puts it this way: "Instead she [Stein] concentrates on feeling (for example, the feeling of joy), since here the 'I' is always present to consciousness. But feeling always requires theoretical acts, and involves values, and both of these lead to the constitution of the rational."[41] In contrast to Baseheart's contention, let me point out that Rudolf Makreel (1996) underlines the basic convergence of Husserl and Stein regarding the cognitive status of feelings: "The phenomenological theory of empathy developed by Husserl and his student Stein gives the feeling involved in empathy not just an emotive quality but a cognitive status."[42] I agree with Makreel in such convergence, for Husserl also underscores the role of feeling, for instance: "I apprehend them [referring to material things of the same type as 'my Body'] as Bodies, that is, I feel by empathy that in them there is Ego-subject."[43] However, this deserves further clarification.

3.3.8 Stream or Sphere of Consciousness

They also share the understanding of a stream or sphere of consciousness. While Stein calls it "stream", Husserl calls it "sphere", although they both use it quite interchangeably. Stein, in constituting the psycho-physical individual, begins with the I as subject of experience, a reflection which reveals that such experience is not isolated, "but set against the background of a stream of such experiences," which are all affiliated with "the present, living pure 'I' thus constituting this unity" called the stream of consciousness—also facing (set against) other streams.[44] In considering the sphere of ownness, Husserl characterizes this sphere as a "stream of my pure conscious

---

[39] Mary Catharine Baseheart, *Persons in the World: An Introduction to the Philosophy of Edith Stein* (Netherlands: Kluwer Academic Publishers, 2010), 40.

[40] Stein, *On the Problem of Empathy*, 91–92.

[41] Ibid.

[42] Cf. Rudolf Makkreel, "How is Empathy Related to Understanding?" *Issues in Husserl's Ideas II*, ed. Thomas Nenon and Lester Embree (Netherlands: Kluwer Academic Publishers, 1996), 200. Cf. Husserl, *Ideas II*.

[43] Husserl, *Ideas II*, sec. 45, p. 172.

[44] Stein, *On the Problem of Empathy*, 38–39.

processes and unities constituted by their actualities and potentialities," all inseparable from my ego.[45] In such stream of consciousness or experience, the world of the I is first constituted, then the world of the other. Such a "world" is essentially intersubjective, regardless of the temporal or logical order of constitutional analysis. Bello (1991) brings out an insight on Husserl's and Stein's understanding of being, where she finds a major point of convergence and divergence between Husserl and Stein: "For Stein, essences, or better still the essentialities, have an existence; ontology roots itself not only in questions of essentialities but also in questions of being."[46] According to her, Stein's "uncovering of the sense of being," with the use of phenomenological method, led her ultimately to deliver "original results." Bello is referring to Stein's (2002) *Finite and Eternal Being* wherein she describes the eternal as possessing the fullness of being. It is noteworthy that, for Stein, "the fullness of sense is contained in eternal being, and eternal being cannot attain its own sense, a sense that fills every creature, except itself."[47] Hence, the divergence considered here is in view of Stein's later works.

3.3.9 The Analogical Character of Empathy

Husserl develops and capitalizes more on the act of *analogizing* that occurs in empathy. For instance, the alter ego (as particular man), according to Husserl, is not perceived as such (as particular man) by me (as ego). Rather, I perceive it as an animate organism in the form of the body "over there", which serves as my motivational basis for apprehending that body as another animate organism similar to me: "I apprehend them as Bodies, i.e., I feel by empathy that in them there is Ego-subject." A certain assimilative perception happens, but not inference from analogy.[48] By virtue of similarity, the body "over there" is connected to my body and assimilated thereof—whereby I apprehend that body as another animate organism. Husserl calls it "analogizing apprehension" wherein "analogizing transfer" (apperceptive transfer) happens—the original sense being transferred to a new non-original (indirect) case (such as someone else) "to the extent that there is givenness beforehand."[49] Such analogizing apprehension, whereby a body is apprehended as likewise

[45] Husserl, *Cartesian Meditations*, sec. 42, p. 90.

[46] But here she compares Husserl and Stein considering already Stein's *Finite and Eternal Being*, which is beyond the scope of this study. Cf. Bello, 274.

[47] Ibid.

[48] Husserl, *Cartesian Meditations*, sec. 50, p. 111. This seems to be where Stein diverges from Husserl, according to Hutt, since Stein advocates "identification" rather than "association" as characteristic of empathy. However, Zahavi argues that in Husserl such identification is in a way necessarily constituted: "The *transcendence* of the object is only constituted at the moment in which its identity across different acts is ascertained. The identification of the object is precisely the noetic correlate of the identity of the object." Cf. *Husserl and Transcendental Phenomenology*, trans. Elizabeth Behnke (Athens: Ohio University Press, 2001), 30.

[49] Ibid., p. 110–111.

an animate organism similar to my own, reveals two conditions: 1) the presented or given, that is, the ego (primally instituted original), is always livingly present; and 2) the appresented, the alter ego, can never attain actual presence. Thus, for Husserl, ego and alter ego, by circumstance, are always and necessarily given "in an original pairing." Husserl's consistent use of pairing, which seems absent in Stein's articulation, suggests a divergence of understanding between them. Stein finds that the theory of association, as basis of pairing, "does not suffice as a principle to explain empathy."[50]

For Husserl, *analogizing* is indispensably crucial in empathy. But to a certain extent, Stein shares Husserl's understanding of the analogizing perception that occurs in empathy, since they both contend against inference from analogy as the kind of analogizing occurring in empathy. According to this sense, as developed by Husserl, the act of empathy essentially consists in the process of appresentation whereby the alter ego is perceived by the Ego as its *analogue*; that is, as similar to itself and yet also distinct (Ricœur, 1967; p.124).[51] This fundamental similarity eventually becomes the basis for any difference that the other has as perceived apart or different from the ego. All this shows the analogical character of the act of empathy, and thereby the analogical character of any form of human encounter that we have.

3.3.10 The Intersubjective Character of Empathy

Both Husserl and Stein constitute intersubjectivity as it is given at the very outset of constituting the stream of ownness. They have both established that empathy is the condition for the possibility of the "intersubjective," given initially by the very act of empathizing. Stein, for instance, constitutes intersubjectivity in the following manner: "Were I imprisoned within the boundaries of my individuality, I could not go beyond 'the world as it appears to me'. At least it would be conceivable that the possibility is demonstrated as soon as I cross these boundaries by the help of empathy [...] Thus empathy as the basis of intersubjective experience becomes the condition of possible knowledge of the existing outer world, as Husserl, and also Royce present it."[52] Husserl, after indicating the intersubjective character of the alter ego and by virtue of what he calls communalization, brings the constitutive analysis of the phenomenological constitution of intersubjectivity: "The transcendental intersubjectivity has an *intersubjective* sphere of ownness, in which it constitutes the Objective world; and thus, as the transcendental 'We', it is a subjectivity for this world

[50] Stein, *On the Problem of Empathy*, 24.

[51] Paul Ricoeur simply refers to this *analogizing apperception* as "the analogical grasping of the Other." He points out how important this is, being at the center of the Fifth Meditation: "This theory of analogical apprehension is, therefore, of considerable weight [...] The encroachment beyond the sphere of ownness is to be accounted for, while at the same time the primacy of originary experience of the ego is affirmed." *Husserl: An Analysis of His Phenomenology* (Evanston: Northwestern University Press, 1967), 124.

[52] Stein, *On the Problem of Empathy*, 64.

and also for the world of men, which is the form in which it has made itself Objectively actual. If, however, intersubjective sphere of ownness and Objective world are to be distinguished here, nevertheless, when I as ego take my stand on the basis of the intersubjectivity constituted from sources within my own essence, I can recognize that the Objective world does not, in the proper sense, *transcend* that sphere of that sphere's own intersubjective essence, but rather inheres in it as an 'immanent' transcendency."[53] In other words, empathy is fundamental—a crucial step—in constituting intersubjectivity (Sawicki, 1997). At the same time, it is empathy that leads to the constitution of "intersubjective objectivity" for things in general, and in particular for human beings (Carr, 1987; Smith, 2003). The intersubjective bursts forth in the very act of experiencing, in the subject and content of experience. The closer we look, by and through empathy, the more it manifests—so that our own self, our experiencing, and the world we experience are all characteristically intersubjective. Phenomenology does usher us towards clarification of intersubjectivity, but it is itself intersubjective (Schnell, 2010). Empathy reveals the *intersubjective* character of the human encounter. It seems that experiencing someone else precisely as other is possible only in an intersubjective experience which the Ego finds inevitably in its experience of itself. The Other is in a way revealed to the Ego in the latter's experience of itself (Hutt, 2009). This reveals an intersubjective phenomenon. It brings to mind the truth-value of the popular saying, "No man is an island." In his last works, Husserl increasingly came to see intersubjectivity as a "dimension or structure that characterizes our intentional life at its very core" (Bell, 1991). This brings out all the more the intersubjective dimension of every conscious individual.

3.3.11 Immanency and Transcendency in Empathy

Husserl and Stein both maintain the distance between the I and the Other, thus establishing the *immanency of the I* and the *transcendency of the Other*—this seems very explicit in Husserl's *Cartesian Meditations* compared to Stein's constitutional explication in *On the Problem of Empathy*. In other words, I obtain the other without me losing myself, and without the other losing itself. The distance between me and the other may have been bridged via empathy, but it is never lost. Empathy maintains the balance between the immanency of the Ego and the transcendency of the objective world as explicated by the Alter Ego (Byers, 2002; Zahavi, 2001). The immanency of the Ego was not overemphasized so as to give way to the transcendence of the Alter Ego, or vice versa. Rather, it strikes a balance, with the result that both are indispensable and serve to complement each other towards a tenable understanding of ourselves and our surrounding world. One thing quite certain about Husserl's theory of empathy is this: in the act of empathy, I maintain my immanence while the other maintains its transcendence. This means that while I maintain my identity as an Ego, the Alter Ego also maintains his/her otherness (Hutt, 2009). This is a crucial

[53] Husserl, *Cartesian Meditations*, sec. 49, p. 107.

point that Husserl, along with Stein, establishes in the overcoming of solipsism (Quito, 2001). It is also important to point out that even in the Ego's sphere of ownness, a certain transcendence takes place, as evidenced by our capacity for self-reflection. Our capacity for *immanent* and *transcendent* acts (of meaning-making, for example) reveals to us the original dualistic structure of the self, i.e., immanence and transcendence—referred to by Armando Rigobello as "primordial belonging and transcendental extraneity" (Rigobello, 1991).[54]

3.3.12 The Reflexive Character of the Act of Empathy

The intersubjective character of empathy appears to be most visible and prominent between human persons. In such encounters, we can also see the reflexive character of empathy, in terms of our knowledge of oneself and others. Since empathy functions within the lived meanings or intentionalities (interiority) of the monadic egos, and since such lived meanings inhere precisely in the human egos, it therefore shows itself to be most prominent and operative in the realm of two or more human beings. Both Stein's and Husserl's accounts of empathy apparently point to another human subject as the empathized one—to be known and understood in his interiority. But empathy points in the first place to self-knowledge and self-understanding on the part of the empathizing subject. Empathy is a journey of understanding towards the other person in his interiority via my own interiority. That is why my understanding of the other always presupposes my understanding of myself. Such understanding, however, brings to light not only our similarity with the other person, but also our difference. Our repeated acts of empathy, then, allow us to be continually enriched in our understanding of one another in our similarities and differences. One fundamental consequence of this is the possibility of establishing mutual understanding and respect between two human beings, among members of a human community or society, and among different societies throughout the world—in our *sameness and difference*. Sawicki points out that, in contrast to Husserl's view, for Stein it is "difference in sameness" that is required for *Einfuhlung* (Sawicki, 1997).

There is something quite paradoxical in this act of empathy—that is, I am able to understand the other person because I have referred his lived meanings to my own. In other words, in reaching out to another person, I must go back to myself, in my interiority, in order to make sense of what I have gathered from the other person in his interiority. In so doing, I am able to reach out to another person in my understanding of him, but I am also able to understand myself better through my encounter with him. A constant encounter with another human person therefore aids us towards *self-examination* and *self-awareness*. It makes sense that Husserl ends his *Cartesian Meditations* (1977) with the following note: "The Delphic motto, 'Know thyself!'

[54] For a detailed discussion, see Rigobello, "A Variation on 'Reduction Within Reduction': 'Interior Extraneity'," *Analecta Husserliana, Vol. XXXV*, ed. A-T. Tymieniecka (Netherlands: Kluwer Academic Publishers, 1991), 259–269.

has gained a new signification. Positive science is a science lost in the world. I must lose the world by epoche, in order to regain it by a universal self-examination. '*Noli foras ire*,' says Augustine, '*in te redi, in interior homine habitat veritas*' – which means: 'Do not wish to go out; go back into yourself. Truth dwells in the inner man'."

3.3.13 The Prominence of Empathy

Empathy is integral and indispensable to Husserl's doctrine of intersubjectivity in particular, and to transcendental phenomenology in general. From the explications of Stein's and Husserl's accounts of empathy, it is evident how the transcendental theory of empathy plays an indispensable role in the phenomenological project in general. Empathy is indispensable insofar as it provides the grounds for the theory of intersubjectivity, which in turn is also indispensable to warrant the legitimacy of an objective world, and accordingly, the legitimacy of objective sciences. Since empathy is at the heart of our encounter with the world around us, any objectivity that we can arrive at is made possible by this act of empathy through its continual process of perception and verification, which confers scientific legitimacy on the empathized objects or subjects and the knowledge acquired thereof. In *Ideas II*, Husserl has seen—through his constitutive analyses—that empathy leads "to the constitution of the intersubjective Objectivity of the thing and consequently also that of man."[55] The theory of empathy, therefore, is instrumental in order for Husserl to arrive at his aim of firmly establishing a *presuppositionless* philosophy and an ultimate grounding for scientific inquiry.

3.3.14 The ethical Dimension of Husserl's and Stein's Constitution of Empathy

Both Husserl and Stein explicitly established the firm grounding for *ethical relation* (Sullivan, 2006, pp. 80–81). The implications of intersubjectivity necessarily oblige us to consider the ethical dimension in our encounter with the other, and more so in our relationships with one another as belonging to the same human community (Son, 1972, p. 128). John Sullivan offers an interesting insight regarding Husserl's and Stein's common quest for truth as leading towards an ethical direction: "Along with Husserl Edith sought after a truth that in later terms would be called 'holistic' for its ability to embrace more than abstract formulae and to include consideration of values, emotions, and ultimate concerns such as peace and solidarity in the struggle

[55] Husserl, *Ideas II*, sec. 47, 178.

for harmony and progress."[56] Both thinkers' pursuit of truth led them towards such an ethical direction. The ethical implication of Husserl's and Stein's constitution of empathy seems inevitable. I find it at its most flourishing in the context of their clarification of the sense of personhood and of community. John Drummond cites, for instance, how Husserl makes clear that "if we are to renew ethical community, we must first renew the ethical individual."[57] Empathy is relevant to our human encounter in the hope of achieving mutual and reciprocal understanding with the rest of humanity, and thereby attaining that solidarity which could bring harmony and peace in the world (John, 1996).[58]

Both Stein's and Husserl's analyses develop a view of the human person that could serve as solid grounding for ethical relation. "The person," according to Husserl (*Ideas II,* 1989), is "*subject of a surrounding world* [...] a person who represents, feels, evaluates, strives, acts and who, in every such personal act, stands in relation to something, to objects in his *surrounding world.*" That is why Stein (1989, p. 116) remarks how important it is ethically that we see and relate with ourselves and with others *as persons* rather as selves: "If we take the self as the standard, we lock ourselves into the prison of our individuality. Others become riddles for us, or still worse, we remodel them into our image and so falsify historical truth." Baseheart (1991) notes that for Stein "only the person as spirit can go beyond the self and relate cognitively and affectively to others in the full sense of these relations."[59] Through empathy (empathic or empathetic acts), we as personal subjects are aware of the *us* for whom the world is a *shared* world, a common world of a *shared* experience. That is the foundation of community that is set forth by both Husserl and Stein in their constitution of empathy.[60]

Empathy is relevant to our human encounter in the hope of achieving mutual and reciprocal understanding with the rest of humanity. In this world of ours—damaged to a great extent by lack of mutual respect, by gross indifference to the plight of the marginalized (less privileged or unprivileged) in our society, by the merciless attack against our humanity—empathy is all the more relevant, not merely as a philosophical theory to enlighten us, but as a way of life to guide our human relationships, and any human encounter for that matter. Empathy presupposes and aims at mutual and reciprocal openness between two or more human subjects. Such mutuality and

---

[56] John Sullivan, "Some Instances of Edith Stein's Humor and Compassion," *Contemplating Edith Stein*, ed. Joyce Avrech Berkman (Notre Dame, Indiana: University of Notre Dame Press, 2006), 80–81.

[57] John J. Drummond, "The 'Spiritual' World: The Personal, the Social, and the Communal," *Issues in Husserl's Ideas II,* ed. Thomas Nenon and Lester Embree (Netherlands: Kluwer Academic Publishers, 1996), 237–253.

[58] Drummond points out how in *Ideas II* Husserl presents "extensive analyses of the person as well as shorter treatments of the notion of community." 238.

[59] Baseheart also points out how Stein, in *Psychology and Humanities*, continued "in her search for a deeper knowledge of person." 51.

[60] In her summary presentation of Stein's constitution of empathy, Baseheart points out that for Stein "the understanding of the 'I' as person and the understanding of person is aided by descriptive analyses of empathy."

reciprocity, a necessary offshoot to Stein's and Husserl's understanding, find an echo in the words of David Bell (1991): "As long as I remain in an exclusively egocentric perspective, genuine reciprocity is impossible: there can only be one space, one ego, one living body, and one intentional world: *mine.* It is only insofar as I can remove myself [...] that there is created so much as the possibility of another ego, say or of an objective space which contains me and other things on an equal footing." If we can, in a way, put ourselves *in the shoes* of others in the most profound sense, then mutual understanding, human solidarity and harmony among us would not be far-fetched, but a dream worth chasing—simply because it is possible and realizable (Hutt, 2009). That is made possible in the first place by our acts of empathy. Our world would be a better place for everyone by developing cultures based on our deep understanding of one another—made possible by empathy.

3.4 An Attempt at Integration

This study is a product of constant reflection, not only on Stein's and Husserl's insights on empathy, but also a reflection on my "lived experiences" as I went through such "excruciating" experience in order to gain clarity. Borrowing the words of Stein, I do not presume to have gained "final clarity" on the matter, considering how tentative my findings are. Nonetheless, I rest content in whatever fruit my arduous efforts have achieved, and remain determined to pursue my investigation to gain more clarity. Let me share some insights I have learned along the way.

3.4.1 Four Salient Points to Consider

Before any concrete synthesis or attempt at integration are made, it is important to set forth specific as well as broader contexts surrounding Stein's and Husserl's constitution of empathy (Payo, 2013). Here are four points to consider.

First, it is clear that Husserl and Stein have their unique contribution to the phenomenological understanding of empathy, and that they have influenced each other's understanding along the way. Having said that, we cannot ignore the intermingling lives and texts of Husserl and Stein which make it extraordinarily difficult to delineate exactly and adequately Stein's unique contribution from that of Husserl's. This difficulty is particularly apparent with the text of *Ideas II*, which bears not only Husserl's understanding, but also that of Stein. Thus, relying on studies focused on this particular text, the present study mainly attributes the account of empathy in *Ideas II* to Husserl, except in certain points where Stein's ideas are inserted or integrated as recent studies have shown. The latest revision of *Ideas II* (2025) is yet to be considered to update and enrich this study. Considering more comprehensively issues inherent in *Ideas II* as well as in other works by Husserl and Stein is a task deemed necessary for further and more definitive findings.

Second, the different ways (and other goals perhaps, personalist or epistemological) of constituting empathy, inevitably affecting the constitution of intersubjectivity, consequently lead Husserl and Stein in different directions. While Husserl's goal or direction is strictly *epistemological,* Stein's direction—aside from being epistemological—tends *towards a more integral understanding of the human person* (Bell, 1991).[61] These differences and their respective understanding of empathy also lead them in different directions in terms of ethical implications. Both Husserl and Stein largely base their approach to ethics on maintaining the balance between identification (similarity) and alterity (difference). That is a matter of emphasis and focus, however, because identity and alterity are twin aspects, inseparably connected, of our encounter with the other through empathy.

Third, the limitations of this investigation (limited both in terms of the materials considered and also the availability of directly relevant materials, especially of primary sources) make the results of this study tentative. Further investigation—for verification, expansion and deepening—is definitely needed. Thus, even the very attempt at integration, which immediately follows our consideration here, is tentative in the aforementioned sense.

Fourth, and most importantly, let it be clear, before attempting to integrate their accounts: that *we have found HUSSERL AND STEIN CONVERGE MORE THAN DIVERGE IN THEIR CONSTITUTION OF EMPATHY as embodied in their specific texts* (i.e., Stein's *On the Problem of Empathy* and *Philosophy of Psychology and the Humanities*, and Husserl's *Cartesian Meditations* and *Ideas II*). In view of this, there seem to be legitimate grounds for attempting a tentative synthesis or integration of their accounts.

3.4.2 A Husserl-Stein Account of Empathy

"Putting oneself in the shoes of another." Let me apply this analogy, if only to creatively bring forth an integration or synthesis of Stein's and Husserl's accounts of empathy as phenomenologically understood. Taken literally, I can put my feet in the shoes of another on the following conditions: (1) the presence and availability of the I and the Other (literally, that I have feet and there's another set of feet); (2) the availability of the shoes and their transferability (condition and capacity to be worn by another set of feet); and (3) the supposition of difference in terms of context (location and situation) of the two sets of feet, the feet of the I and the feet of the Other.

Such self-evident conditions demonstrate the necessity of two separate subjects (hence, two separate sets of feet, and at least a pair of shoes worn by one of the two subjects) and the ability (along with inclination) on my part to put my feet (myself) "in the shoes" of another subject. Furthermore, in view of the conditions of the shoes

---

[61] Bello remarks how Husserl "felt compelled to perfect his method as an **essentially cognitive instrument**." See "Ontology, Metaphysics, and Life in Edith Stein," 272.

being made available for me to wear, there is the condition that the shoes are not being worn by the other in order for me to wear them; for my feet and his feet cannot occupy the same space at the same time. In view of my capacity to wear the other's shoes, considering they have been made available for me to wear, there is also the condition of actualizing or realizing my capacity; i.e., I must act or move in order to make it happen. For even if I am capable and the shoes are available, I still would not be wearing those shoes if I did not make a move to go and wear them. In fact, it may not really be necessary for me to actually wear the shoes. I can have a pretty good idea of the other's experience in wearing his shoes by comparing or referring back to my own experience of wearing my shoes. The other's experience of wearing his shoes remains his experience, but the sense of his experience (now objectified) is being transferred to me through my act of referring his experience back to mine (comparing or analogizing). Hence, such sense combines with my actual experience. That somehow describes to us an act of empathizing or what empathy is.

But, of course, this saying, "putting oneself in the shoes of another," is a figurative and analogical one, properly speaking. In common understanding, it means *putting oneself in the context or situation of another person in order to somehow experience his experience, and thereby understand him better*. We use empathy to describe this act of understanding another person in his particular context or situation. In applying the said analogy, I have indicated elements of an empathic act, namely: (1) the presence and availability of the self and the other person, along with the semblance of similarity in both of them; (2) the availability of the other perspective and its transferability; (3) the intuitive insight of the difference between the two subjects in terms of context and point of view; and (4) the self experiencing the world as the other experiences it, or the other experiencing the world as the self experiences it (as portrayed in the very act of wearing the shoes on the part of the other). These elements point out broadly the salient aspects in both Stein's and Husserl's accounts of empathy.

Delving deeper into such elements, we find the following aspects to be crucial in constituting empathy, for both Stein and Husserl: (1) the Pure I or Ego; (2) the givenness of the I to itself and of the world to the I; (3) the point of orientation, whereby the I takes the center of orientation, and a shift of orientation results in an empathic understanding of the other and the rest of outer reality; (4) the outer world (the empathized object/subject) or reality as correlate of the subject of experience (the empathizing subject); (5) the importance of expression as the vehicle towards objectification and interpretation via empathic reflection; and (6) the understanding of values in the realm of motivation.

Moreover, it is clear how Stein and Husserl converge in the following areas of their explication: (1) the I and the Other as correlative (two separate subjects); (2) the double givenness (two-sidedness of empathy) of an empathic act—that is, as given non-primordially and primordially; (3) the inviolable distance between the I and the Other, which leaves no room for a complete fusion or unity of the I and the Other; (4) the important role of sensations, as well as voluntary movement, as constituents of the living body; (5) the notion of the stream of consciousness; (6) the notion of a zero-point of orientation, resulting in the I's being at the center of orientation, and allowing a possible shift of orientation from the I to the Other—seen as the condition of the possibility of empathy, thus very important in constituting the foreign individual; and (7) the fundamental distinction between the realm of nature and realm of the spirit. The conglomeration of these and many other aspects in their clarifications or constitutional analyses serves to ground us in a penetrating and illuminating understanding of their insights into the issues of phenomenology and intersubjectivity in general, and more so of empathy in particular. Considering these crucial and indispensable elements, we may attempt to formulate or articulate an integrative account—that is, an integration of two accounts which converge and diverge to a greater or lesser degree.

What I have done here is to map out areas only; that is, paving the way towards an integration. Strictly speaking, I have been hesitant to formulate a synthesis of Husserl's and Stein's accounts. I thereby take these considerations as merely preparatory to a *creative* formulation of such an integration. After all the clarifications considered in the course of my investigation, it is clear that Stein's and Husserl's early constitutions of empathy do not radically diverge from each other. Rather, they converge largely, especially in the fundamental and essential elements of the phenomenological theory of empathy. Thus, this work for me is open-ended, certainly a beginning towards further clarification, elaboration and deepening of the insights already gained vis-à-vis the insights found in the later works of Stein and Husserl, certainly with the invaluable help of studies on Stein and Husserl.

3.4.3 Tentativity of the Results Gained

It is clear that Husserl and Stein each have their unique contribution to the phenomenological understanding of empathy, and they influenced each other's understanding along the way. The intermingling lives and texts of Husserl and Stein make it extraordinarily difficult to delineate exactly and adequately Stein's unique contribution from Husserl's. This may be because their constitution of empathy, considering the texts investigated, converges more rather than diverges; Stein herself explicitly admits as much. Thus, the consideration of other works by Stein and Husserl is necessary for further and more definitive findings. But in view of some divergences pointed out, I can say that the different ways of constituting empathy—inevitably affecting the constitution of intersubjectivity—consequently led Husserl and Stein in different directions. While Husserl's goal or direction was strictly epistemological, Stein's direction was towards a more integral understanding of the human person (Estafia, 2004, 2011). Such differences in direction and their respective understanding of empathy led them also on different paths in terms of ethical implications.

These and other findings I have arrived at, and even the attempt at integration of Husserl's and Stein's viewpoints, are characterized by tentativity. The previously mentioned limitations of this investigation (the limited materials considered and also the availability of directly relevant materials, especially of primary sources) make this unavoidable. I have been conclusive at certain points in certain areas, but overall, the results of this study remain largely tentative. I am referring concretely to the broader spectrum of intersubjectivity and phenomenology, towards which empathy is a step. Yet we can be conclusive in this, insofar as the role or significance of empathy as a step is concerned: that it is necessary and indispensable in the work of constitution—that is, in constituting reality through and in our consciousness. That is why it is so crucial, not only for Stein, but also for Husserl and for every phenomenologist or seeker of truth for that matter.

More specifically, the results of this study are tentative in view of the original works of Stein and Husserl thus considered. First of all, the texts themselves are not here available for my consideration. But even if they were made available, my non-proficiency in the German language and the shorthand which Husserl used in his original manuscripts prevents me from taking a closer look at these invaluable and irreplaceable materials. Thanks go to W. Stein, M. Sawicki, M. C. Baseheart, D. Cairns, R. Rojcewicz and A. Schuwer, among others, for their invaluable translations have made this study possible. Nonetheless, with my deep appreciation for their scholarly and arduous work, my work (insofar as it is based on their translations of Stein's and Husserl's works) is not conclusive insofar as it needs further verification in the original sources.

Furthermore, my study has dwelt largely, though not exclusively, on Stein's *On the Problem of Empathy* and Husserl's *Cartesian Meditations*, particularly the *Fifth Meditation.* That fact alone makes this study and its results tentative in view of the other equally (and even more) important works by Stein and Husserl.[62] Noteworthy of these are Stein's *Philosophy of Psychology and the Humanities*, which I have dealt with cursorily, and Husserl's *Ideas II* and other works on intersubjectivity. Taking account more closely of these works, particularly the latest edition of Husserl's *Ideas II* (2021), would yield a more conclusive understanding of Stein's and Husserl's accounts of empathy, intersubjectivity and the whole phenomenological enterprise.

3.5 Final Remarks

First, our study of empathy brings to our consideration the relationship between Stein and Husserl, which we see as one of fidelity as well as independence. While Stein remains faithful to Husserl's phenomenological way, she also exhibits independence of thought in her phenomenological analyses and insights, particularly with regard to the problem of empathy.

Secondly, in view of such relationship between Stein and Husserl, this study highlights their work of collaboration in the pursuit of phenomenology. This is shown by their respective texts (i.e., Husserl's *Ideas II* and Stein's *Philosophy of Psychology and the Humanities*) and lifelong correspondence.

---

[62] Antonio Calcagno, for example, explicitly considers Stein's *On the Problem of Empathy* to have been written under the direction of Husserl. Thus, her later works certainly show her independence from such explicit influence. Cf. *The Philosophy of Edith Stein* (2007).

Significantly, the phenomenological account of empathy underscores the distinct role of empathy in the philosophical quest for truth. Empathy highlights the phenomenological way of seeing the world as integral or holistic. We usually find empathy outside the philosophical discourse (e.g., in the arts and psychology). Stein's and Husserl's phenomenological accounts of empathy made empathy distinctly as an indispensable issue in the philosophical quest for knowledge and truth.

Let me now wrap up this brief account by sharing with you my attempt to make a kind of personal integration at the end of this research work. It was a very challenging experience to gain clarity. Edith Stein describes it as "excruciating," yet in the end deeply fulfilling. My findings are largely tentative, but one thing is clear: that empathy is crucial to intersubjectivity in particular and to phenomenology in general, thus indispensable for any phenomenologist to consider. And having considered Stein's and Husserl's accounts of empathy, the explicit way towards integration of their accounts is perhaps at the most basic phenomenological understanding of empathy.

As a priest for 22 years, who became acquainted with the phenomenological understanding of empathy, I would confidently say that *empathy is a way to love.* St. John XXIII has struck me as conveying in his action and words the authentic sense of empathy—that is, by bridging the gaps that separate us from one another. Pope Francis, of blessed memory, had been doing the same thing in his papacy; that is why we were often amazed and surprised by his "reaching out" to others, especially to those in need of mercy and love. I realize that *empathy*, in its broadest sense, extends beyond philosophical discourse. Empathy opens the way to a whole range of human understanding. Most importantly, the depth and breadth of understanding which empathy could usher us paves the way for love, compassion and solidarity among us. I hope and pray that our consideration of empathy is never confined to philosophical discourse, but goes far beyond—in the overflowing richness of what we call "lived world", where we may plunge ourselves in the immense dimensions of empathy, in our personal living encounter with one another, in this life and beyond.

References

Baseheart, M. C. (1991). *Person in the World: Introduction to the.* Philosophy of Edith Stein Kluwer.

Bell, D. (1991). *Husserl.* Routledge.

Berkman, J. A. (Ed.). (2006). *Contemplating Edith Stein.* University of Notre Dame Press.

Byers, D. (2002). *Intentionality and transcendence: Closure and openness in Husserl's phenomenology.* University of Wisconsin Press/Noesis Press, Ltd..

Calcagno, A. (2007). *The philosophy of Edith Stein.* Duquesne University Press.

Carr, D. (1987). *Interpreting Husserl: Critical and comparative studies.* Martinus Nijhoff.

Estafia, J. C. (2004). *Edith Stein on the problem of empathy: Towards being human.* MA Thesis in Philosophy. University of Santo Tomas.

Estafia, J. C. (2011). *Edith Stein: Her contribution to the dialogue between faith and reason.* Doctoral Dissertation. Faculty of Philosophy, Ecclesiastical Faculties, University of Santo Tomas.

Herbstrith, W. (1992). *Edith Stein: A biography.* (B. Bonowitz, trans.). Ignatius Press.

Hernandez, M. R. (2008). *The concept of intersubjectivity in the personalism of Edith Stein.* MA Thesis in Philosophy. University of Santo Tomas.

Hutt, C. (2009). Identity, alterity, and ethics in the work of Husserl and his religious students: Stein and Levinas. *Philos Today, 53*(1), 12–33.
John J. Drummond. (1996). The 'Spiritual' world: The personal, the social, and the communal. In In Nenon, T. & Embree, L. (Eds.), *Issues in Husserl's ideas II* (237–253). Kluwer Academic Publishers.
Lebech, M. (2008). Stein's phenomenology of the body: The constitution of the human being between description of experience and social construction. *Maynooth Philos Pap, 5*(2008), 16–20.
Makkreel, R. A. (1996). How is empathy related to understanding? In Nenon, T. & Embree, L. (Eds.), *Issues in Husserl's ideas II*. Springer Netherlands.
Nenon, T. (Ed.). (1996). *Issues in Husserl's "Ideas II,"*. Kluwer.
Payo, F. (2013). *Edith Stein and Edmund Husserl on empathy: Towards a comparative synthesis.* Doctoral Dissertation. University of Santo Tomas.
Quito, E. S. (2001). *Phenomenology: Edmund Husserl and Edith Stein.* De La Salle University Press.
Ricœur, P. (1967). *Husserl: An analysis of his phenomenology.* Northwestern University Press.
Sawicki, M. (1997). *Body, text, and science: The literacy of investigative practices and the phenomenology of Edith Stein.* Kluwer.
Schnell, A. (2010). Intersubjectivity in Husserl's work. *META: Res in Herm, Phen, and Pract. Philosophy, II*(1), 9–32.
Smith, A. D. (2003). *Routledge guidebook to Husserl and the cartesian meditations.* Routledge.
Son, B. H. (1972). *Science and person: A study on the idea of "Philosophy as Rigorous Science" in Kant and Husserl.* Van Gorcum.
Stein, E. (1989). *On the problem of empathy, 3rd revised edition (W. Stein, Trans.).* CWES III, ICS Publications.
Stein, E. (2002). *Finite and eternal being: An attempt at an ascent to the meaning of being.* (K. F. Reinhardt, Trans.). CWES IX. ICS Publications.
Sullivan, J. (2006). *Edith Stein: Essential writings.* Orbis Books.
Sweet, W., & Feist, R. (Eds.). (2003). *Husserl and Stein.* The Council for Research in Values and Philosophy.
Tymieniecka, A.-T. (Ed.). (1991). *Husserlian phenomenology in a new key: Intersubjectivity, ethos, the societal sphere, human encounter, pathos.* AHua. The Yearbook of Phenomenological Research XXXV. Kluwer.
Tymieniecka, A.-T. (Ed.). (2002). *Phenomenology world-wide foundations–expanding dynamics–life-engagements: A guide for research and study.* Kluwer.
Ullrich, M. (1996). Nature and spirit. In Nenon, T. & Embree, L. (Eds.), *Issues in Husserl's ideas II*. Kluwer Academic Publishers.
Zahavi, D. (2001). *Husserl and transcendental intersubjectivity.* (E. A. Behnke, Trans.). Ohio University Press.
Zahavi, D. (2010). Empathy, embodiment and interpersonal understanding: From Lipps to Schutz. *Inquiry, 53*(3), 285–306.

Fr Francis Bayhon Payo is a priest for 23 years from the Archdiocese of Capiz, Philippines. He earned a licentiate-masteral degree in Sacred Theology and a doctorate in Philosophy (study on "Edith Stein and Edmund Husserl on Empathy: Towards a Comparative Synthesis") at the University of Santo Tomas (UST), Manila. He taught at UST, St. Paul's University (Manila) and Colegio de la Purisima Concepcion (Capiz). He also taught in Sancta Maria Mater et Regina Seminarium (Capiz) and four other seminaries. He is currently pursuing a doctorate in theology (focusing on Stein) at the UST Faculty of Theology while helping Fr. Joeffrey Brian Catuiran (parish priest) at Sta. Rosa de Lima Parish, Diocese of Pasig.

Part II
Intersubjectivity and Social Ontology

Chapter 4
Personally Embedded Qualitative Social Ontology: Issues from Stein and Husserl

Francesca De Vecchi

Abstract In this study, I deal with Husserl's and Stein's social ontology with reference to Husserl's and Stein's works in the years 1917–1925, in particular Husserl's *Ideas II* and Stein's *Beiträge* and their intricate intellectual authorship. I point out some of the differences between Husserl's and Stein's gaze on the social landscape, but I also claim that both focus on the relationship between, on the one hand, personal identity and "personally living", and on the other, personal collectives and "socially living". How does being part of a certain type of collective contribute to individuating the person I am? How may the *personal being* felicitously meet the *social being*? These are some of the crucial questions that Husserl's and Stein's phenomenology of the person introduces into social ontology, in making it a personally embedded qualitative social ontology.

Keywords Stein and Husserl · Qualitative social ontology · Personalistic attitude · Personal-common-world · Social wholes · Personal wholes

4.1 Introduction

4.1.1 Outline

In my paper, I work on Husserl's and Stein's social ontology, along with its roots in their phenomenology of the person and intersubjectivity. I suggest that Husserl and Stein are both founders and prime exponents of what I call "qualitative social ontology"—a phenomenologically embedded social ontology—that combines ontology of the person and ontology of societies and communities and analyses the

F. De Vecchi (✉)
San Raffaele University, Milan, Italy
e-mail: devecchi.francesca@unisr.it

M. Lebech (ed.), *Stein and Husserl's Phenomenology of the Social in Ideas II*, Women in the History of Philosophy and Sciences 25,
https://doi.org/10.1007/978-3-032-17339-3_4

so-called social world from a "personalistic attitude" (Husserl, 1912–1928/1952).[1] Indeed, a qualitative social ontology has developed within phenomenology from its beginnings, and Husserl and Stein are among its protagonists, together with other phenomenologists like Max Scheler, Adolf Reinach and Gerda Walther.[2] The expression "social ontology" itself was coined by Husserl, and appears for the first time in his manuscript *Soziale Ontologie und deskriptive Soziologie* (1910, in Husserl, 1905–1920: 99–104). Many of Husserl's works after the *Logical Investigations* (Husserl, 1901) present a social-ontological *Leitmotiv* and similarly Stein's output following her dissertation on empathy (Stein, 1917) is marked by a deep attention for the ontology of sociality. Of all Husserl's and Stein's works, Husserl's *Ideas II* is one of the most intriguing and relevant texts to an inquiry into the phenomenologically-embedded social ontology of Husserl and Stein: it focuses extensively on social-ontological issues and most of its manuscripts (posthumously published as *Ideas II*) were collected and edited by Stein, as is well known. I must point out that a new edition of Husserl's *Ideas II* has recently been published (2025), in which Stein's arrangement of Husserl's manuscripts has been deeply modified. Hence, there are now two different editions of *Ideas II*: an "old one" (Husserl 1912–1928/1952) and a "new one" (1908–1930/2025). In my paper, I focus on the old one because it best illustrates the interconnections between Husserl and Stein.

Therefore, in my paper I refer to Husserl's *Ideas II* (Hereforth, by *Ideas II* I mean only the "old edition": Husserl, 1912–1928/1952), and more precisely to the third section on the constitution of the so-called "spiritual world" (*geistige Welt*). Moreover, I discuss some arguments presented by Husserl in his manuscripts on the *Phenomenology of Intersubjectivity*, in particular *Gemeingeist I–II* (Husserl, 1921; 1918–1921), and in his Kaizo articles on Europe and its renewal (Husserl, 1923–1924). On Stein's side, I refer to her *The Problem of Empathy* (Stein, 1917), *Individual and Community* (Stein, 1922) and *An Investigation Concerning the State* (Stein, 1925). These texts seem to me the most significant when working on the social ontology of Husserl and Stein during the period 1917–1925, which is the one that interests us, according to the target of the present volume.

4.1.2 Claims

Firstly, I suggest that Stein's and Husserl's social ontology share the same perspective: that of a personally embedded social ontology, so to speak. Indeed, the "personalistic attitude" that characterizes their social ontology allows us to account for our everyday life-experience of the world as a personal one. Under the personalistic

[1] I displayed the idea of "qualitative social ontology" as an ontology of the person and society together in De Vecchi (2022). I coined and used the expression "qualitative social ontology" for the first time in De Vecchi (2016), and subsequently developed my account of a qualitative social ontology phenomenologically embedded in several texts: De Vecchi (2017, 2020, 2022, 2024, 2025).

[2] For an overview of the contribution of phenomenology to social ontology see Salice and Schmid (2016), Moran and Szanto (2016), and Zahavi (2025).

attitude, persons *live personally*: as subjects of personal acts, i.e., of various degrees of motivationally connected position-takings; and as subjects who experience things of the world as value-things, i.e., as various types of goods and ills, which are the correlates of persons' motivational lives and with respect to which persons may take positions in their life. Under the "personalistic attitude", things are value-things of a personal world that is always a common and shared world: a social world, to which we as persons belong both as individuals and as parts of various forms of collectives. Moreover, under the personalistic attitude, things of the world are value-things independently of whether they are natural or artefactual: this is a crucial point that differentiates Husserl's and Stein's qualitative social ontology from "mainstream social ontology", rooted in the analytic philosophy tradition, which separates natural and artefactual kinds, social world and natural world, social facts and natural facts (e.g., Searle, 2010).

Secondly, I point out some differences between the social ontology of Husserl and that of Stein. I argue that Stein zooms in on a freeze-frame, so to speak, of our social landscape, a still image whose protagonists are already individuated and structured persons who are in relation to other persons: there is me, there is you, there is she/he, and there are I–you relations, and I–you/–she/–he relations, which may constitute different types of collective we. Stein's phenomenological analysis begins with the given-data of persons who already exist as such, and, in a fine-grained way, deals with different types of collectives or wholes—masses, societies, communities, states—that are characterized by various degrees of individualizing force upon their individual members and by different kinds of individuals' intentionality—empathy acts, social acts, emotional contagion, experiences of identification. This posture characterizes Stein's *The Problem of Empathy* (1917), *Individual and Community* (1922), and*An Investigation Concerning the State* (1925).

On the other hand, Husserl's gaze adopts a wide-angle lens through which he attempts to capture a dynamic sequence of scenes of our social landscape, whose protagonists are personal individuals and collectives filmed as characters of a world-horizon in which they become persons. In other words, Husserl focuses on the conditions of possibility for becoming persons who live personally. In *Ideas II,* being a person is essentially connected with being part of collectives that are the subjective poles of a "common surrounding world" that, in turn, is the motivational horizon with respect to which persons can live as subjects of personal acts. In other texts meanwhile, such as the manuscripts *Gemeingeist I–II* (1921, 1918–1921) and the *Kaizo* articles (1923–1924), Husserl's social-ontological landscape changes and becomes more dynamic: besides the personal common world, in which individuals live personally and in a position-taking modality, there is also a less personal social world, in which individuals live together in more gregarious ways and in a taking-for-granted modality with respect to themselves and the world. Husserl addresses there the issue of the renewal as the possibility of assuming (again) a personalistic attitude and become agents of position-takings.

More generally, my itinerary through Husserl's and Stein's works aims to raise a fundamental issue about the relationship between persons and society: that between, on the one hand, personal identity and "personal living", and on the other, the various forms of collectives and "social living" of which individual persons may become subjects. How may the fact of being part of a certain type of collective contribute towards individualizing the person who I am? How may it influence, negatively or positively, my personal flourishing? Conversely: how may individual persons constitute a social collective as the kind of collective it is? How can the personal being meet, felicitously or not, the social being? These are crucial questions that both Husserl's and Stein's phenomenology deal with and introduce in social ontology; naturally, each in their own way.

4.2 Personal Embedded Social Ontology

The personalistic attitude is the natural and spontaneous attitude in which we live in the world together with others; this is the attitude proper to the qualitative social ontology of Husserl and Stein. This is one of the basic theses I want to articulate in this paper. In the personalistic attitude, things are given to us qualitatively; the world is a world for me and for us, a world "on hand" (Husserl, 1913, § 27), mine and ours, in which we live personally, in a perspective-taking modality. It is a "common surrounding world" of goods and ills, in which value- and disvalue-qualities are realized and to which *I* individually and *we* collectively respond.

4.2.1 Phenomenology of the Personal and of the Social

Husserl's and Stein's qualitative social ontology deals with both the being of persons and the being of things; both the subjective and objective poles of the intentional relations that define the world. Indeed, it is no accident that the project of the phenomenology of the world as a "spiritual world", i.e., as a "personal world" (*geistige Welt*), as presented in Husserl's manuscripts that were collected and edited by Stein in *Ideas II*, is based precisely on an analysis of both the subjective and objective poles of this personal world: an analysis of the personal identities, both singular and collective, and of the things of the world that constitute the correlates of the intentional, and more precisely, motivational life of the persons—of their feeling, thinking, willing and acting. This is one of the great insights of *Ideas II*: a phenomenology of the world and of the things of the world as things that have meaning and relevance for us, as things that count for our existence, can only be at the same time a phenomenology of the personal subjects who live in that world, who are influenced and motivated by it, who are themselves part of that world, and who at the same time create and modify it; in short, of the subjects who have to do with the world personally, as subjects of position-takings and responses in regard to that world. Therefore, the phenomenology of the world in which we live daily cannot but be at the same time a phenomenology of the subjects who live in this world as persons.

This is clearly stated in the opening of the third section of *Ideas II*, dedicated precisely to the constitution of the so-called "spiritual world" that is the "personal world". Indeed, the German *geistig* is a complex word, including as it does both the sense of "spiritual" and the sense of "personal". The "spiritual" is the "supernatural", which is detached from the sphere of the physical or material, and also the "mental" as separated from the body; but from the phenomenological perspective, "spiritual" means specifically "personal", and precisely the being of the person that is not reducible to the physical-material level on which it is nevertheless grounded. Thus, the term *Geist*, "spirit" and "mind", and the adjective *geistig*, "spiritual" and "mental", in phenomenology mean "person" and "personal" and properly indicate a subject in which the classical separation between spirit and mind on the one hand, and matter and body on the other, is overcome. Husserl points out that this perspective involves a new foundation of psychology and, more generally, of all human sciences: "What is needed is a new "psychology" of an essentially different type, a general science of the spirit that is neither "psychophysical" nor naturalistic"; i.e., a new general science of the person that does not reduce the person to their psychophysical basis and that deals with person and world as a whole, in overcoming the classic "opposition between *nature and the world of the spirit*, between the natural sciences and the human sciences, the sciences of the spirit; between a natural scientific theory of the psyche on the one hand and the theory of personhood (of the ego, egology) as well as theory of society (theory of community) on the other hand" (Husserl, 1912–1928/1952, En. Tr. 1989: § 48).

This is the project of a theory of the person that includes a theory of the personal world and its social formations: a social ontology based on an ontology of the person, "a theory of the society in person", we could say. There is no doubt that Stein shares this project: not only did she edit most of Husserl's manuscripts collected as *Ideas II*, but she also explicitly endorses this project in her *Beiträge zur philosophiscen Begründung der Psychologie und der Geisteswissenschaften* (Stein, 1922).

Indeed, as I have already mentioned *Ideas II* is a collection of Husserl's manuscripts, most of which Edith Stein had worked on during the years of her assistantship (1916–1918), although Husserl never gave his *placet* for their publication. The new edition of *Ideas II* (Husserl, 1908–1930/2025) arranges and integrates Husserl's manuscripts differently.[3] With respect to the perspective of the qualitative social ontology, which I am illustrating, as rooted in a phenomenology of the person that implies a phenomenology of the world as a personal one, and vice versa, my line of argument does not change: it seems to me that, with the *Teilentwürfe 2* focusing on the "constitution of the personal world", the new edition of *Ideas II* confirms

[3] More precisely, the manuscripts of the new edition of *Ideas II* (Husserl, 1908–1930/2025) have been collected and structured according to their belonging to three main fascicles: the *Teilentwürfe 1* (1912), *2* (1913), and *3* (1915) which occupy the first part of the volume. The second part of the volume consists of the temporally successive *Ausarbeitungen* relating to some of the topics addressed in the *Teilentwürfe*. The second part of the volume also includes the *Nachwort* that Husserl had published in 1930 in the *Jahrbuch*. In the new edition, the *Third Book* of *Ideas II* (which had been introduced as such by Stein) has been eliminated; the manuscripts that were part of it are now placed in the first part of the volume, as they belong to *Teilentwurf 1* (1912).

Husserl's general project of a phenomenology that is a social ontology. Nevertheless, there is a point that is worth specifying: the distinction between the genuinely Husserlian perspective and that attributable more to Stein, as editor of the Husserlian manuscripts, relates to Husserl's analysis, now brought out well in the new edition, of personal habitualities and passivities and their role in the constitution of the personal being, both individual and collective. The new edition would show that, according to Husserl, there is a deep self that is intertwined with, but not identical to, the deciding and positional self that is the core of the person as the author of their living. In other words, individual and collective persons are subjects of various kinds of experiences that may be more or less personal; i.e., more or less engaging as subjects of position-takings. Corresponding to these different degrees of personally living, there may also be social worlds which are less personal, and types of collectives that individuals may become parts of *via* contagion, conformism and other forms of passive experiences.[4] This is why, I think, Husserl's personal and social world changes and becomes more complicated than it is in the first or "old" edition of *Ideas II*. Other texts (the already mentioned manuscripts on the *Phenomenology of intersubjectivity*, in particular, *Gemeingeist I–II*, and the *Kaizo* articles) also include deeply-rooted but unwitting and passive instances of the personal identities, as I shall show.

Concerning Stein's endorsement of the phenomenological project of *Ideas* II, she writes in the foreword of her *Beiträge*:

> the following investigations undertake to penetrate into the essence of the psychic reality and of the person (*Geist*). [...] The first of the two following investigations attempts to work out plainly the twofold basic lawfulness – causality and motivation – operating together within one psychic subject with a sensuous-mental (*sinnlich-geistig*) essence. The second investigation broadens the consideration from the isolated psychic individual to super-individual realities, and thereby seeks to attain broader insights into the composition of the personal (*geistig*) cosmos. (Stein, 1922, En. Trans. 2000: § *Foreword*; I have modified the English translation of *Geist* and *geistig*)

The project behind Stein's *Beiträge* is the same as that of Husserl's *Ideas II*: they deal with the specific nature of persons as subjects both of causally grounded experiences and of motivationally founded personal acts; moreover, they focus also on the social, inter- and super-individual life of the persons as subjects who together constitute collectives of individuals. Stein's first *Beitrag* intends to show that the "motivation" is the proper and specific lawfulness of the life of persons and is irreducible to the category of causality that also defines the being of persons as psychophysical subjects. The second *Beitrag* works on the different forms of collectives and the various ways in which personal individuals can constitute these collectives, as subjects of both causal and motivational connections.

In the *Foreword*, Stein herself speaks of the relation of her investigations to the work of Husserl. She mentions her "helping Professor Husserl for nearly two years with the preparation of large publications" and Husserl's "important influences" on her own work: "during this time, all his manuscripts from the last ten years have

[4] About Husserl's analysis of instances of habituality and passivity and their role in the personal identity in the new edition of *Ideas II* (Husserl, 1908–1930/2025), see Caminada (2019, Ch. 11).

been at my disposal (among them those that have to do with the topic of psychology and the humanities as well)". A large part of these manuscripts are precisely those edited and collected by Stein for the publication of Husserl's *Ideas II*. Therefore, not only is what Husserl writes in *Ideas II* marked by Stein's editorial work on his manuscripts, but what Stein writes in her *Beiträge* is also significantly influenced by Husserl's thought and the editorial materials on which Stein worked, so much so that Stein affirms: "I was not sure whether I would have to regard something as my own research result or as an internal appropriation of transferred thought motifs" (Stein, 1922, En. Tr. 2000: § *Foreword*).

Notwithstanding this whole intricate intellectual relationship between Stein and Husserl concerning *Ideas II*, and the major attention that Husserl pays (in the new edition of *Ideas II* and also in some manuscripts of the *Phenomenology of Intersubjectivity* and in the *Kaizo* papers) to the role of passive and habitual instances for the constitution of personal identity, intersubjectivity and social world, it seems to me that both Stein and Husserl agree on the fundamental lines of what a phenomenology of the person should be. According to them, the phenomenology of the person is a phenomenology of human beings described not only as psychophysical and causally characterized individuals, but also as subjects of personal acts (be they cognitive, affective, or practical acts); and at the same time it is a phenomenology of the world not only as a world of stimuli and causes which determine us but also as world of things of value or disvalue, a world of goods and ills, to which we respond motivationally. A world that is never made up solely of single individuals, but of collectives of individuals. Hence it is a phenomenology of the person that involves a phenomenology of persons together, as collective subjects of a world shared by them: a phenomenology of the person that is, again, a phenomenology of society.

4.2.2 Value-Things and Relation of Motivation in the Personal World

Husserl's *Ideas II* highlights the ontological varieties of the entities that inhabit our personal and surrounding world. These are objectivities that may be natural or artefactual entities, social and institutional entities on the one hand, and subjectivities that are personal subjects, both individual and collective, on the other.

> This surrounding world is comprised not of mere things but of use-Objects (clothes, utensils, guns, tools), works of art, literary products, instruments for religious and judicial activities (seals, official ornaments, coronation insignia, ecclesiastical symbols, etc.). And it is comprised not only of individual persons, but the persons are instead members of communities, members of personal unities of higher order […]. (Husserl, 1912–1928/1952, En. Tr. 1989: § 49; Husserl, 1908–1930/2025, *Teilentwurf 2*, § 2e)

In the existential posture that is specific to the personalistic attitude, these different kinds of entities (which are in the surrounding world and constitute it, both the objectivities and the subjectivities) are understood in their personal significance and relevance, as cultural entities, and not in their mere materiality or as mere nature—"mere things"—as is, by contrast, required by the naturalistic attitude.

The objectivities are the correlates of our motivational life. As persons we refer to, create and use them as private, public and institutional objects in our everyday lives: artifacts, work of arts, and social, institutional, religious instruments, which we share in social practices of various kinds. Analogously, subjectivities are also not experienced as merely psycho-physical beings, but rather as personal subjects who are both individual and collective persons: communities and personal unities of a higher order that transcend the individual persons on which they are founded. In the personal world,

> Instead of the causal relation between things and human beings as natural realities, there is substituted the *relation of motivation* between persons and things, and these things are not the things of nature, existing in themselves – i.e. the things of exact natural science with the determinations claimed there to be only Objectively true ones – but are the experienced, thought or in some other way intended and posited things as such, intentional objects of personal consciousness. (Husserl, 1912–1928/1952, En. Tr. 1989: § 50; Husserl, 1908–1930/2025, *Beilage VII* zu § 3a: 239)

Therefore, both the objectivities and subjectivities of the personal world are ontologically qualitative entities that we experience in their relevance and salience with respect to our own lives. The personal world is never just "pure nature", but is also and above all a cultural world whose entities have a human and personal relevance, and this independently of whether they are natural or artefactual.[5] Indeed, natural as well as artefactual entities can be cultural. The personal world is the world that is inhabited by any kind of entities having a personal relevance, in the sense that they are crucial for human beings to exist as personal subjects. Thus, in this perspective, even natural entities have a personal and cultural relevance, since they also are the motivational correlate of our intentional experiences. That plant, that fruit, that animal being, etc. (and not only that work of art, that artefact, that institution) is an entity overflowing with personal significations and value-qualities—be they practical, aesthetical, moral, religious, institutional, etc. We perceive the value-qualities of the entities as essential parts of their identity; i.e., as entities that are more or less useful, pleasant, flourishing, beautiful, holy, etc.[6] We do not experience the entities of the personal world just as "existing in themselves", devoid of value-qualities and meaningless, as merely material bodies that are neutrally there, as only occupying a space–time position.

In this sense, the phenomenological view of the world as a personal world opens up a qualitative social ontology, whose domain is not limited to entities created

---

[5] On the belief that only artefactual and not also natural entities can be cultural and on the common tendency "to confuse the cultural with artefactual", see Embree (2003: 40–41).

[6] On the common surrounding world as a "value-laden and meaning-structured lifeworld", see Kohàk (2003: 25–26). On the surrounding world as a cultural world and its relationship with the personalistic attitude, see Melle (1996: 24): "The life-world is a personal-cultural world. Culture in the broad sense comprises all realities, which have predicates of meaning. 'Culture in general is thus the correlate of performing subjectivity'" (quoting Husserl, 1919: Ms. FI 35/115b).

by human beings, but also includes natural entities.[7] So, with respect to classical ontology and contemporary social ontology, *Ideas II* makes an extremely original and fruitful point. There is no social world as opposed to a natural world, as there is in the perspective of analytical social ontology; in "motivational and personal living", indeed, there is no separation between the causally determined physical-psychic level and the transcendent level of being agents and authors in their own lives. From this perspective, natural and artefactual objects, entities whose existence is supposedly independent of the intentionality of individuals and entities whose existence is instead created as well as maintained in being by the intentionality of individuals, are in both cases all entities that are part of our world as a humanly and personally meaningful world, rich in qualities of value that are existential saliences for us.[8]

Our acts of experiencing things are always also acts of value-perception: acts of valuing the things we are experiencing and acts of perceiving their value-qualities.

> In [acts of valuing] the object is brought to consciousness as valuable, pleasant, beautiful, etc., and indeed this happens in various ways, e.g. in original givenness. In that case, there is built, upon the substratum of mere intuitive representing, an evaluating which, if we presuppose it, plays in the immediacy of its lively motivation, the role of a value- "perception" (*Wert-Wahrnehmung*) […] in which the value character itself is given in original intuition. (Husserl, 1912–1928/1952, En. Tr. 1989: § 50)

In acts of value-perception (also called value-feeling acts), we experience value-qualities as originally given data that constitute the qualitative layer of the ontological structure of the things we encounter in the world, and it is this qualitative layer that makes the things into the goods and the ills of our motivational life as personal subjects.[9]

Along the same lines, Stein affirms in her *Beiträge*:

> A value constitution goes hand in hand with every object constitution. Every fully constituted object is simultaneously a value-object. Basically, the value-free world of mere things is an abstraction […] (Stein, 1922: 144, En. Tr. 2000: § 2.c) *Affective acts*, 160)

Value-qualities question the persons who experience them: to feel value-qualities is to be engaged in a relationship in which we are called to respond to them, to take a position towards them, and thus to act and live personally: feeling emotions, making

---

[7] A paradigmatic example of the classical way of understanding "social ontology" as the domain of only social entities, meant as entities that, unlike natural entities, are created by human intentionality and language, is presented by Searle (1995, 2010).

[8] As paradigmatic examples of analytically driven social ontology separating "brute facts," which are natural and "ontologically objective," (i.e., existentially independent of the intentionality of individuals), and "social and institutional facts," which are "ontologically subjective" (i.e., existentially dependent on the intentionality of individuals), see Searle (2010) and Gilbert (2013). Gilbert demarcates the boundaries of the social world to those only of joint commitment, excluding from it anything that, while not the object of joint commitments, is nevertheless shared and experienced together qualitatively.

[9] The most exhaustive phenomenological account of value-perception and value-qualities is that of Scheler (1913–1926), which constitutes a real treatise on axiology.

judgments, and making decisions related to them.[10] This is precisely why things as value-things are the motivational correlates of our being persons, subjects of a personal life. Stein explains this very well:

> beauty is not like the sensible qualities of the object to which beauty adheres. Those qualities place no claim upon me. I can perceive them, but I need not trouble myself further about them. Basically, they don't matter to me. Beauty, on the other hand, insists that I inwardly open myself to it and let my inner self be determined by it. And for as long as this inner contact is not effected, for as long as I withhold the response which beauty requires, beauty doesn't entirely divulge itself to me. The intention inhering in the mere information (*Kenntnisnahme*) [of feeling value] remains unfulfilled. (Stein, 1922: 159, En. Tr. 2000: § 2.c) *Affective acts*, 143)

Beauty as value-quality is distinguished from sensible qualities by its motivating force toward the person who experiences it: value-qualities have an existential relevance for us that sensible qualities do not have.

So, Husserl and Stein's phenomenology of the personalistic attitude is a personally embedded qualitative social ontology, according to which our experience of things is always also an experience of their value-qualities, for in the personal world things are never given to us as merely material but also as valuable things. Qualitative social ontology is a phenomenology of the person and a phenomenology of the world, social and cultural, which mutually imply each other: it is a phenomenology of the person and society together, i.e., a phenomenology of the world as a personal, social and cultural world.

4.3 Intersubjectivity, Personal Identity and Different Types of Collectives

Both Stein and Husserl modify the concept of "intersubjectivity" in enlarging it progressively, from the I-you relation to the collective we. Indeed, Stein begins with empathy acts understood as face-to-face encounters between an I and a you (Stein, 1917), then focuses on collective experiences by which the you-other makes way for the members of communities and other kinds of collectives (Stein, 1922), and finally turns her attention to social acts and in particular law-making acts of the citizens of a state (Stein, 1925). Analogously, Husserl expands the concept of intersubjectivity in *Ideas II*: from unilateral and bilateral acts of empathy to acts of communication and mutual agreement, and to forms of collective intentionality. All these kinds of intentional experiences constitute the "common surrounding world" whose subjective poles are both individual persons and collective persons (Husserl, 1912–1928/1952). Both Husserl and Stein masterfully analyse the variety of intentionality-types

[10] On the relationship between the experience of values and emotions in phenomenology, and more generally on personal responses to the experience of values, see Stein (1917, 1922), Scheler (1913–1926), Hildebrand (1922), Vendrell Ferran (2022), De Monticelli (2019, 2021) and Cimino et al. (2022).

through which an individual may turn toward another and, more generally, have something to do with others. This is what I call "heterotropic intentionality", including the intersubjective, collective and social intentionalities (De Vecchi, 2014), by which individual persons live and manage intersubjective, interpersonal and social-ruled relations, and also constitute forms of collectives.

However, besides this common enlargement of the concept of intersubjectivity, I suggest that Stein and Husserl outline two different socio-ontological landscapes in adopting different perspectives on them. Husserl focuses not only on the various kinds of relations among persons and the different kinds of collectives they create, but also and above all on the horizon that allows individual and collective persons to exist as such; i.e., on the conditions of possibility for their being subjects of personal acts (see, *infra*, §§ 4.3.2–4.3.2.5). Stein, on the other hand, does not address this issue. In her account, the horizon does not figure as a possibility condition for becoming persons; persons exist already and always as subjects of personal acts. Rather, Stein concentrates on the relationship between deployment of individual personalities and their being parts of certain kinds of collectives, between their personal identity and the social roles they may assume in societies or communities, between personal types and social types, and correlatively she addresses the issue of whether and how individual persons can make of a collective a personal one (see, *infra*, §§ 4.3.1–4.3.1.3).

4.3.1 Stein's Social Landscape

According to Stein (1917), acts of empathy are the ones that found human relationships: in empathizing, I experience the other as such, as a subject of experiences of different kinds (both psycho-physical and personal). I may have a more or less defined comprehension of those experiences (indeed, there are degrees and stages of empathic acts), which are non-originally given to me: those experiences belong to the other, not to me, even though I can perceive them. So, acts of empathy constitute the relationship between an I and a you, a relation in which I do not take the place of the other, but rather, in remaining myself I see the other as such and grasp her experiences (Stein, 1917, En. Tr. 1964: § 2 *Description of empathy in comparison with other acts*).[11]

Then, in widening her phenomenological gaze on human relationships, Stein focuses on the "we": the starting point of her analysis is always an I and the intentional experiences pertaining to the I as related to another, the I-you relationship between two individuals, and never an already existing we (as for instance we can find in Scheler's social ontology; see Scheler 1913–1923, 1913–1926).

Stein's investigation in *Individual and community* "attempts to get to the bottom of the question of how it is possible to have a community as a higher-level subject and a community life" (Stein, 1922, En. Tr. 2000: § Introduction).

[11] On Stein's empathy account, see Zahavi (2014), Moran and Szanto (2016) and De Vecchi (2019).

> To start with, we ought to investigate *which* individual experiences are suitable for helping to build up a communal experience (*Gemeinschaftserlebnis*), because not everything that surfaces in the current of consciousness is useful for this. Experiences with an individual sense, which means experiences that target something that has significance only for the individual alone, remain confined to that individual and play no constitutive role beyond him or her. (Stein, 1922, En. Tr. 2000: § 2. *Components of the experiential current*)

The point of departure is always the individual person and their intentional experiences: the ones that are intersubjective, social and collective, i.e., the ones that are not reducible to solitary or singular experiences. Here Stein adopts the same criterium as Reinach (1913, § 3) and Scheler (1913–1926, § *ad 4. Einzelperson und Gesamtperson*) in distinguishing between, on the one hand, experiences that involve only a single individual and, on the other, experiences that, to be performed, need necessarily to involve other individuals.

At the origin of the community, and more in general of any other type of collective life, are always and only single individuals; according to Stein, first there are individuals, and then communities and other forms of collectives that are constituted by various sorts of heterotropic individuals' experiences. Thus, at the origin of the social world there are only individuals who may create collectives, and never the other way around: forms of collectives from which individuals may emerge.[12]

> The community is founded essentially in individuals. Its character changes if the individuals belonging to it change their characters, or if new ones enter and old ones withdraw. [...] we pursue the "origin" of the community and seek out its rooting within the individual. (Stein, 1922: 215, En. Tr. 2000: § 4. *The relationship of foundation of individual and community*, 238)

Very interestingly, Stein deals with the relation between individuals and collectives in considering it as a parts-whole and a whole-parts relation. Even though Stein does not make it explicit, she avails herself here of Husserl's theory of wholes and parts, according to which wholes are "unitarian foundations" (*einheitliche Fundierung*): unlike mere aggregates or sums, wholes are characterized by existential-dependence relationships that qualitatively inter-permeate the being of the wholes and of their parts (1901, *III Logical Investigation*, §§ 7, 10 21).[13]

As Stein affirms in the above-quoted passage, the character of the collective-whole depends on the characters of the individuals-parts who constitute it; it changes if they change or if new individuals become parts of the whole and others cease to be parts of it. In other words, the qualitative being of the collective-whole modifies according to the being of its individuals-parts—in Husserl's terms, the nature of the wholes depends on the nature of their parts (Husserl, 1901, *III Logical Investigation*, § 7, 10).

---

[12] This opposite way is the one of Scheler (1913–1923; 1913–1926): in the beginning there is the social unity of individuals who live experience of fusion and identification (*Einsfühlung*).

[13] I cannot adequately develop here the issue of how, and how fruitfully both Stein and Scheler apply Husserl's "unitarian foundation" account in their social ontology. I addressed this extraordinarily important issue for a qualitative, phenomenologically embedded social ontology in De Vecchi (2022). Here, in this paper, I limit myself to making a couple of points on this issue.

Then Stein's ontological analysis of collectives goes further: she claims that there are different ways of being individuals-parts of a collective-whole and different ways of being a collective-whole in relation to its individuals-parts. This qualitatively ontological perspective allows us to distinguish different modalities in which parts-persons and wholes-collectives may be ontologically connected (i.e. different kinds of "unitarian foundations"), and to account for the various types of collectives that we experience in our lives. Stein argues that there are qualitatively different types of wholes-collectives according to the stronger or weaker individualization activity that the whole-collective operates on its parts-persons, and according to the intensity of the bonds—more or less deep or superficial, strict or flexible—that bind the persons as parts of a certain whole-collective.

The different types of collectives identified by Stein include the society (e.g. business companies), the community (e.g. families), the mass (e.g. the crowd of fans at the stadium) and other "mixed modes" of collectives that present traits of different collective types (Stein, 1922, En. Tr. § 4.e) *Mixed modes of social unities*, 283).[14]

4.3.1.1 Society and Community

Let us focus on Stein's distinction between the societies and the communities.

> The life of the society subsists in the functionality required of its members by its purposes. For that, it is non-essential that precisely these individuals perform the work in question. In principle, anybody is replaceable by someone else. On the other hand, life as a member of the association does not coincide with someone's total (*gesamt*) life but forms only a small sector of it. Insofar as the individuals fit the configuration of a society (fulfill their functions), they become representatives of certain types (worker, supervisor, head of the party, and so forth). [...] these types are rooted in the composition of the society and not in the composition of the individual personality (*individuelle Persönlichkeit*). [...]
>
> A general personal type (which is presupposed so that the individual can fit into any association at all) corresponds to the association (or the purposive union) overall [...] (Stein, 1922: 230–232, En. Tr. 2000: § 4.c), *Individual and society*, 256–257)

Stein points out very clearly that the relationship between the parts-persons and the whole-society does not involve a strong individuation of the parts-persons, who simply have to fulfill the social roles needed to make a society function. A society is a "mechanism", created by individuals' social acts and collective acts of cooperation, which aims at the realization of a scope that is the *raison d'être* of the society (Stein, 1922, § 4c. *Individual and Society*). What is relevant are the roles or functions of the society, and these can be assumed by anybody who is able to accomplish them. The persons-parts of the society-whole are "social types" (Stein, 1922, § *The social types*) that impact on the being of the persons-parts only superficially; they merely label and fix the persons in social functions that do not engage the individual being of the persons and occupy just a marginal and superficial portion of their life. These

[14] Stein refers to, but also originally revises it, the account of Tönnies (1887) that contrasts society and community (Stein, 1922, § *Introduction*).

roles, indeed, are not rooted in the being of the persons-parts but in the being of the society-whole, which consequently operates a weak individuation activity on the being of the persons-parts.

On the contrary,

> [the community] is based in the distinctiveness (*Eigenart*) of the individuals who enter into the community, so are all organs and functionings that build up the community determined the same way. (Stein, 1922: 236, En. Tr. 2000: § 4.d), *Individual and community*, 261)

Unlike society, the individuation force of the community-whole on its persons-parts is strong. Indeed, the community-whole requests that its persons-parts engage themselves personally, and penetrates the being of its persons-parts: in order to belong to the community-whole, the persons-parts must allow the community to enter deeply into their lives. On the other hand, the community-whole depends on the personal individuality of its parts: its persons-parts have to be just the persons they are. In other words, the community-whole founds itself on the unicity and essential individuality of its persons-parts. The community grows and dies like an "organism" that is nourished by the individuals who live in it, in a communality of life (*Gemeinsamkeit des Lebens*) grounded on the sharing of various kind of experiences, which may be more or less personal—i.e., more or less free and aware depending on the individualizing force and on the request of personal engagement exercised by the community (Stein, 1922, En. Tr. 2000: § 4.d)cc) *The genesis of community…*).[15]

Here, Stein again applies Husserl's "unitarian foundation" account (1901, *III Logical Investigation* § 21), according to which there may be a double wholes-parts constitution: not only are wholes constituted by their parts, on which they are grounded, but in turn they may also constitute their parts in defining and individuating them. This is the case of wholes that Husserl refers to as the "second sort" (Husserl, 1901, § 21): wholes that are typically "emergent wholes" as opposed to those that remain "immanent wholes".[16] Emergent wholes are those whose parts are not "pre-destinated to be parts" (Husserl, 1901, IIIRL, § 7) of those wholes of which they are parts, and they may exist also outside those wholes. So, whenever these parts are connected together in a whole, they constitute a whole that is a new type of thing—as the individual persons who are the parts of a whole-collective. On the other hand, "immanent wholes" are the ones whose parts can only exist as parts of those wholes of which they are parts, such as the color and extension that are existentially dependent parts of any material body as a whole.[17] Analogously, Stein specifies that the structure of the community-whole displays a double direction of foundation: there

[15] Stein does not endorse Scheler's distinction between the community as "life-community" (*Lebensgemeinschaft*) and the community as "common person" or even "collective person" (*Gesamtperson*) (Scheler, 1913–1926); rather, she argues for just one type of community within which different degrees of personal living may be realized (Stein, 1922, § 4.d)ee) *Various types of community*).

[16] The expression "emergent wholes" is by Conni (2005). On the double constitution that characterizes "emergent wholes", see Caminada (2022) and De Vecchi (2022, ch. 4)., where I specifically worked on collectives as emergent wholes.

[17] I dealt extensively with this issue in De Vecchi (2022).

is not only the bottom-up foundation, from persons-parts to the community-whole, but also a top-down foundation from the community-whole to the persons-parts, and this top-down foundation individualizes the being of the persons-parts.

> We'll have to take a closer look at this "reaction" (*Rückwirkung*) of the communal life to the individual life, which is to be observed in all spheres which are affected by communal life. (Stein, 1922: 141, En. Tr. 2000: § 2.b) *Categorial acts*, 157)

The fact of being part of a certain community-whole profoundly modifies the being of the persons-parts: in becoming a part of that community-whole (for example, the Italian Parliament or the *Filarmonica della Scala Orchestra*), I am now a new and different person, since the role I assumed in those communities-wholes impacts strongly on my personal identity.

Stein also observes that there are collectives in which individuals are definitively not individualized; this is the case of the mass, which is the most anonymous and impersonal form of collective. Its individuals-parts live only experiences of "contagion" and "suggestion" as dispositions to be inculcated with "groundless" emotions and beliefs, disengaged therefore from a motivational connection, and confuse the one with the other (Stein, 1922, En. Tr. 2000: §4.b) *Individual and mass: mass contagion*).

4.3.1.2 Personal Types and Social Types

Stein's analysis of the social-ontological landscape is always fruitful for qualitative social ontology, since it is very fine-grained and attentive to the dynamic structure of social collectives and their relationship with individuals' personal identity. In particular, Stein's analysis of the social roles, social types and personal types is extremely valuable and original for a further reason: because it is never dogmatic. Stein observes that certain social roles "are not 'masks' that the individual takes up and under which the individual conceals his 'true face'"; rather they "correspond to one or another trait of the individual essence", for "the typical behavior and the type itself receive their individual imprint (*individuelles Gepräge*) from the persons who enter into them" (Stein, 1922: 265–266, En. Tr. 2000: § *The social types*, 293). Stein makes it very clear that the assumption of a role by an individual-part in a certain social whole does not necessarily have a depersonalizing effect. On the contrary, the person-part may fill the role with her personal style, and the assumption of that role may provide her with the opportunity to deploy her individual personality. In this sense, we can speak of the realization of true "personal types" that are to be distinguished from mere "social types".

Stein had already addressed the issue of the "personal types" in the last section of her book on empathy, where she observes that "every single person is already herself a type" (Stein, 1917, En. Tr. 1964: § *Personal types...*, 114), a "personal structure that marks off a range of possibilities of variation within which the person's real distinctiveness [*Ausprägung*] can be developed 'ever according to circumstances'" (Stein, 1917, En. Tr. 1964: § *Soul and person*, 110). According to Stein, there is a "personal kernel" (Stein, 1917, En. Tr. 1964: § *Soul and person*) that "lends an individual stamp" to every trait of the social type that the person exemplifies in the collective to which he/she belongs (Stein, 1922, En. Tr. 2000: § *Character of the community and typical character of the community member*, 263).[18] Indeed, the essence of the individual person is not reducible to the sum of typical traits, those of the collectives-wholes of which she is a part.[19] Stein points out that a good relationship between individual persons and collectives is achieved when the individual person imbues the social type with her personal style, and makes it a personal type; only under this condition can the individual person unfold her personality within the collective. On the other hand, there is the possibility of collective types, such as communities, that may be personal collectives only on condition that they are constituted by individuals who have a genuine personality. As Stein puts very clearly:

> If somebody were "a typical German" in his essence and nothing else, he wouldn't have a genuine personality [...]. For the genuine being of the community has its origin in the personal distinctiveness (*persönliche Eigenart*) of individuals. If there were only typical representatives (that means, representatives who'd be nothing other than types), then any further evolution of the communitarian person (*Gemeingeist*) would be impossible, and then no communitarian person (*Gemeingeist*) would have been able to build up at all. (Stein, 1922, En. Tr. 2000: § 4.d)bb) *Character of the community and typical character of the community member*, 263, slightly modified)

I want to point out that, in this quoted passage, Stein uses the expression *Gemeingeist* to refer to a type of community that is a genuine personal collective. Given the context related to a community type which is here defined as personal, I suggest translating *Gemeingeist* as "communitarian person". However, more generally, *Gemeingeist* is translatable as "communal person" or "common person", and it is therefore a synonym of *Gesamtperson*, which is the expression used by Scheler in his *Formalism* for indicating the qualitatively highest form of social unity, the one in which the deployment of individuals' personality and the constitution of a genuine personal community existentially depend on each other. Indeed, *Gesamtperson* means "common person" or "collective person" (Scheler 1913–1926, § ad 4. *Individual Person and Collective Person*),[20] and it is also semantically close to Husserl's *personale Ganzes* that is translatable as *personal whole*. Besides this constellation of expressions relating to this form of specifically personal collective, the point I also want to make concerns the thickness and ambiguity of the expression

[18] See De Santis (2025) on Stein's account of individual essences in her phenomenology of the person.

[19] This is a very crucial point for our contemporary societies in which the impact of the typical traits of persons on the personal identity is enormously increased by the social profiles on digital platforms to which people's identities seem to be reducible. This is also the issue of "personal digital identities" that, in the current debate, is more commonly addressed in terms of the permanence vs. evanescence of the living body in the virtual or online sociality of the social networks (Fuchs, 2014; Kekki, 2020; Osler, 2024) than in the Steinian terms of the relationship between essential individuality of the person and social types and profiles that persons may endorse.

[20] *Gesamtperson* is translated as *social person* in the English published translation of Scheler (1913–1926).

Gemeingeist; in Stein (1922) it has a "positive meaning". Indeed, it indicates the collective type that has the highest degree of personal fulfilment (and is to some extent synonymous with the Schelerian *Gesamtperson*), while in Husserl, as we will see, it refers to the complex range of possible collective types—which may be more or less personal or impersonal—and has also a "negative meaning", since its extension also includes collective types that are not genuinely personal.

4.3.1.3 Social Wholes and Personal Wholes

Developing Stein's insight about the distinction and, at the same time, the crucial relationship between "social" and "personal" identities in the collectives-wholes and in their persons-parts, I suggest that, within the general set of collectives which I called "social wholes", we pick out those types that are specifically forms of socially living in which *personal being* and *personal living* are realized; that is, those collectives in which the personalities of persons come into play and they fulfill themselves as subjects of personal acts. I propose to call these types of collectives "personal wholes", adopting Husserl's expression *personale Ganzes* (Husserl, 1921/1973b, *Gemeingeist I*, § 9, 182–183). In this way, I claim that, within the general set of social wholes, there are kinds of social wholes that are specifically personal wholes.

In her *Investigation concerning the state* (1925), Stein focuses on the state as a paradigmatic case of "personal whole". According to Stein, the state is a particular kind of community whose distinguishing mark is "sovereignty": the state is sovereign (it is its own master), for it is the "author of its own law" and thus "the subject of law-making acts" (Stein, 1925, I, §§ 1–2). The state is a "personal whole" precisely because it is the agent of personal acts such as law-making acts, which are a particular kind of social act.[21] In the state as a whole, the citizens-parts who assume the role of legislators and who are the agents of law-making acts (e.g., the members of the parliament), exemplify very well the case in which the personal being of the individuals-parts finds in the social being of the community-whole the possibility to flourish: in fulfilling their social role, the legislator fulfills herself as a person. The better legislators accomplish their role in representing the state, and the more effectively they act on behalf of the state, the more they act personally; they individualize themselves and express their personalities as agents of personal acts and position-takings that are, at the same time, public and institutional acts.

In Stein's analysis of the state, the paradigm of the social relationship is always the I-you relation: in this case, the relation existing between the state, the legislator and her law-making acts, on the one hand, and the citizens to whom law-making acts are addressed, on the other. Moreover, Stein deals with the "felicity conditions" of law-making acts for the interpersonal life between citizens and legislator as representing

[21] In her *Investigation Concerning the State*, Stein takes up Reinach's account of social acts as personal acts—according to Reinach, indeed, social acts are "spontaneous", free acts (Reinach 1913)—and develops a social ontology of the state, of the law, and of social acts (see De Vecchi, 2015, 2019; Taieb, 2020).

the state—an interpersonal life that is normative and institutional. Indeed, Stein's account of the law-making acts aims at showing the crucial role that the interpersonal relation between the legislator (who, as an organ of the state, is the agent of the law of the state) and the citizens (who are the addressees of the law and the counterpart of the legislator as an organ of the state) has for the uptake and performance of law-making acts on the one hand, and for their fulfilment on the other. Stein develops a very accurate analysis of what the legislator should do to ensure citizens perceive and recognize the laws of the state. In particular, Stein identifies two types of problems that may obstruct the uptake of legal provisions and that are grounded in the interpersonal gap that characterizes the legislator-citizen relationship: (i) the impersonal moment of legal provisions; and (ii) possible cases of infelicity of the performance of legal provisions.[22] Stein's concern understands the possibility condition for the state to be a genuine community in the sense of a "personal whole", in which person-parts of the social whole can express their personality in affirming themselves as personal agents of their state-whole.

In conclusion, Stein develops a qualitative social ontology that is founded on individual persons. She deals with individuals who already exist as personal subjects; at the beginning of the sociality there is always an I, you, she, he, and never a "we". As I will show below, this is an important different feature of her social ontology in comparison with that of Husserl. Nevertheless, Stein focuses on the ways in which the personal identity of individuals interfaces with social roles and types that individuals may assume in different kinds of collectives, and she also considers the individualizing force—more or less intense and efficacious—that collectives may exercise upon the being of persons: collectives may be more or less personal (or more or less impersonal) depending on how their social roles and types may be personally fulfilled and become an occasion for the deployment of individuals' personality.

4.3.2 *Husserl's Social Landscape*

As already mentioned, it is necessary to distinguish between the social landscape outlined by Husserl in the *Ideas II* manuscripts edited by Stein and the social landscape that emerges from some Husserl's manuscripts on the *Phenomenology of Intersubjectivity* and from some passages of the *Kaizo* articles. In Husserl's *Ideas II*, persons are above all the subjects of position-takings and personal acts in a common-surrounding world experienced in the personalistic attitude, and this is an illuminated and polite social world. In the other texts, by contrast, Husserl gives deeper consideration to the role played by the passive and inertial instances of persons, and consequently the social world and its forms of collectives also assume more obscure traits with respect to individuals' awareness: implicit habituality, conformism and contagion are features that are now considered by Husserl in his analysis of the social landscape.

[22] See Stein (1925, I, §2d). I worked on this issue in De Vecchi (2015, 2017).

4.3.2.1 A Clear and Illuminated Social Landscape

In Husserl's *Ideas II*, persons perform unilateral and bilateral acts of empathy, which found acts of communication and mutual agreement, as well as acts of collective intentionality. Persons are the subjects of intersubjective and social acts through which any individual person understands the experiences of others and communicates with them. It follows that, in performing acts of empathy (both unilateral and mutual) and social acts of communication, singular persons constitute "personal communities" and "personal unities of higher orders" (Husserl, 1912–1928/1952, § 51).

Individually and collectively, persons live and act against a background, a horizon: the common surrounding world that is a personal and social world, experienced in the personalistic attitude, as we have seen. This account of our personal world as a common surrounding world of personal collectives enlightens a crucial feature of the ontological structure of Husserl's social world: there is a personal world only on the condition that it is also a shared world, one whose subjective poles are not only singular persons but also and above all collective persons. So, personal collectives and the common surrounding world are essentially correlates of each other and depend existentially on each other: neither the common-surrounding world, nor personal collectives can exist independently of each other.[23] The common-surrounding world cannot exist without personal collectives as its counterparts and personal collectives cannot exist without a common-surrounding world as their motivational correlate. Let us focus on this point:

> In the *comprehensive experience of the existence of the other*, we thus understand him, without further ado, as a personal subject and thereby as related to Objectivities, ones to which we too are related: the earth and sky, the fields and the woods, the room in which "we" dwell communally, the picture we see, etc. We are in a relation to a common surrounding world – we are in a personal collective: these belong together. We could not be persons for others if a common surrounding world did not stand there for us in a community, in an intentional linkage of our lives. Correlatively spoken, the one is constituted essentially with the other. Each ego can, for himself and for the others, become a person in the normal sense, a person in a personal collective, only if comprehension brings about the relation to a common surrounding world. (Husserl, 1912–1928/1952, En. Tr. 1989: § 51: *The person in personal association*, 201, slightly modified; Husserl 1908–1930/2025, § 3 *Das Wesen der Personalen Subjektivität*, §b *Die Person in Personenverband*, 192)

Husserl claims that in experiencing the other, in intersubjective and social acts, we understand the other as a personal subject and consequently as a subject who is in relation to the same objectivities we are also in relation to. According to Husserl, there is therefore a mutual implication between our perceiving the other as a personal subject and our grasping the other as a subject who refers to the same kinds of entities (i.e., to the same world) that we also refer to. Being a personal subject and being in relation to the same world is the same thing. We understand the other as

[23] See Husserl (1912–1928/1952: § 51), where he develops an analysis of intersubjective and social acts that are fundamental for the constitution of the common-surrounding world and of the collective subjects as essential correlates of the common-surrounding world.

a personal subject only if we and the other share a common world. Correlatively, we understand the other as a personal subject only if the other belongs to the same personal collectives to which we also belong: only if we and the other are members of the same personal collectives.

Now the crucial point here is that the idea of "common surrounding world" entails important implications concerning the ontological status of personal individuals and communities. Indeed, sharing a common-surrounding world and, correlatively, belonging to a personal collective are conditions of possibility for being a person. This implies that from Husserl's perspective (by contrast with Stein's), it is not possible to isolate singular persons as if they existed autonomously, before they constitute personal collectives and the common-surrounding world to which they refer and in which they live. On the contrary, there are personal individuals and collectives because there is at the same time, or already, a common surrounding world which persons as such emerge from and belong to. When we experience the existence of another person, we immediately grasp their being a personal subject of motivational relations with value-things of various kinds, to which we are also related in our common surrounding world. According to this perspective, it is not the case that an I, a you, and she/he exist independently, and then, together, constitute a collective we—as Stein (1922) instead argues.

This is *the personological and correlational thesis of Husserl's social ontology*: there is a mutual implication between our experience of the others as personal subjects and our grasping of them as subjects relating to the same things to which we ourselves relate—the things of the common surrounding world to which we also belong. These things are not mere things of bare physical-material reality, but value-things, goods of various kinds for us—useful and vital goods such as the earth and the sky, the fields and the forests; cultural, aesthetic, and moral goods such as the room in which we dwell together and the picture we contemplate together. Therefore, to be a person means to live personally as a subject who is in relationship with things as goods of a common surrounding world; things that constitute the motivational correlate of our living, things that we use, contemplate, create, modify, etc., in relationship with other personal subjects. We understand others as personal subjects only if they share the things of the common surrounding world—if they grasp their sense of being, the fundamental value-qualities that identify things as a certain kind of good.

Thus, there is also *another personological but more specifically socio-collective thesis* that Husserl argues here, and which clarifies the sense of the previous thesis: we are persons, who as such are in relation to a common surrounding world of goods, only if each of us is part of personal collectives. *Personally living implies socially living; being and acting as a person is founded on being part of a social whole*, even though not all social living implies a personal living in a genuine sense, because *not all forms of social wholes are at the same time personal wholes* in which persons live together as subjects who express their personality in performing personal acts.

Jointly, these two theses—the personological and correlational one and the personological and social-collective one—assert that each person lives personally, and has a personal common world as the motivational correlate of their living, only if she is part of a collective of persons.

To recap, belonging to a common surrounding world and sharing it with others is the condition of possibility for living as a person, in the strong sense of "person"—meaning a subject who unfolds her uniqueness in a motivationally grounded life. This condition, as we have seen, is essentially connected to being part of a personal collective: nobody can be a person on their own. The constitution of each individual person is dependent on the constitution of the collective of persons to which she belongs; I am the person I am insofar as I am part of one or more collectives that constitute the subjective pole of the common surrounding world.

Husserl's ontology of the person is thus essentially connected with his social ontology. From this perspective, therefore, it is not possible to experience persons who live independently of one another, outside personal collectives and the common surrounding world, which is their objective correlate; outside personal collectives and outside the common surrounding world, persons cannot live personally.

4.3.2.2 A More Complicated and Dynamic Social Landscape

In the Husserlian ontological-social landscape of the common surrounding world, frescoed in the pages of *Ideas II* on the "Constitution of the Personal World", there seem to be present only types of collectives that are properly "personal wholes": collectives that are grounded in the personal living of motivational life and position-takings such as acts of empathy, social and communicative acts, and acts of agreement and cooperation. Forms of social unity founded on passive and inertial, unaware and involuntary experiences are not considered there. These are collective types in which persons are the subject of identification, conformism and personal anonymity, as in the gregarious living of the masses or even, in a more nuanced way, in the so-called "life-communities" (*Lebensgemeinschaft*) to which Scheler devotes a fine-grained and profound analysis (Scheler, 1913–1923; 1913–1926).

Contrastingly, in some manuscripts collected in the *Phenomenology of the inter-subjectivity* (Husserl, 1973a, 1973b) and in the *Kaizo* articles (Husserl, 1923–1924), Husserl also considers a personally less evolved, more primitive and anonymous kind of collective: a living together that has "taking for granted" and not personal position-taking as its dominant tonality, and that is pulled by implicit and inertial connections. It is a collective dimension of complicity, based on the mentality of common sense and habitualities that have settled there. In this impersonal or pre-personal dimension of social living, there are just "social wholes" that are not also personal wholes. These are collectives in which both affective and cognitive experiences, such as emotions and beliefs, are not personally experienced (by taking a perspective on them), or even consciously shared among individuals, but are transmitted by means of automatisms to imitation, by "contagion" and "suggestion", devoid of motivational connections and founded on causal processes—just as Stein's accurate analysis of the phenomena of contagion and suggestion shows in her description of the mass as the most anonymous form of collective (Stein, 1922, En. Tr. 2000: §4.b) *The individual and the mass: mass contagion*; see *supra* § 4.3.1.1).

In these other texts, therefore, Husserl's social-ontological landscape changes and grows complicated. The common world is not only a motivational world of value-things with respect to which individuals take positions and affirm themselves as personal agents; it is also a world in which persons may live more passively and inertially, in a taken-for-granted modality.

4.3.2.3 Higher-Order Personalities and De Facto Communities

In the manuscripts *Gemeingeist 1* and *Gemeingeist 2* (Husserl, 1921; 1918–1921), to which Husserl himself gave the title *Gemeingeist*, the focus is on the "common person" (as we can translate *Gemeingeist*) grasped in its ambivalence and ambiguity: as genuine "personal wholes" on the one hand, and as a common mentality, common sense and common attitude that implicitly and passively connect persons together and create forms of collective that are not personal, on the other. Therefore, we find there the theme of personal acts of empathy and communication that—as in the ontological-social landscape of *Ideas II*—found "higher-order personal unities", which Husserl also calls here "higher-order personalities" (*Personalitäten höherer Ordnung*), and among which he cites universities and states. But at the same time, we also find in these texts a background social level, that of "de facto communities (*Wirkungsgemeinschaften*) lacking the unity of an overall collective volition and action": forms of collectives that are not yet or no longer personal and are not subjects of personal acts, i.e., position-takings (Husserl, 1918–1921, §§ 2, 6, my translation).

> [...] one must distinguish between personalities of a higher order (*Personalitäten höherer Ordnung*), genuine personal unities, and mere [...] *de facto* communities (*Wirkungsgemeinschaften*). A language does not come into being as a constitution in the parliamentary state. (Husserl, 1918–21, § 6; my translation)

"*De facto* communities" are, for example, constituted by persons who speak the same mother tongue; these are forms of social unity that exist actually but implicitly, in the sense that personal individuals do not have to take a position toward these kind of communities in order for them to exist and do not have to perform personal acts that renew their creation or revalidate their existence. The language community, in fact, is a widespread and pervasive bond that is not yet a personal community in the proper sense: "a language community is not a bond between persons that creates a personal whole (*ein personales Ganzes*)". And, "speakers of the same mother tongue act by means of an idiom that is recognized in their own linguistic community, but is not a way of personal acting" (Husserl, 1921, § 9: 182–183). It is rather a "linguistic camaraderie" (Husserl 1905–1920/1973: 106) that does not need to be lived personally, as instead it is the case of "higher order personalities", like the state or parliament as the author of the state's constitutional set of rules.

Indeed, individuals who are the parts of the parliament as a whole perform personal social acts: the law-making acts, which create the constitutional set of rules of that state, are free acts and spontaneous position-takings, as Stein shows in her work on the state (Stein, 1925; see *supra*, § 4.3.1.3). To be part of a collective that is the author of the state's constitutional set of rules is to be the subject of a voluntary personal activity that aims at the realization of a legal, institutional and collective work. From this perspective, not only the parliament but also the state in general is a "personality of a higher order" that it is a personal collective in a strong and specific sense. Indeed, individuals live and act in it by performing personal acts as parts of the state-whole, constantly taking a position vis-à-vis it, and in so doing they renew the foundations of the institution that is their own state. At the same time, the state and the parliament as wholes individualize their parts-persons as citizens and parliamentarians and they express themselves as personal beings by performing personal acts proper to citizens and parliamentarians.

A state is thus a "personality of a higher order" precisely insofar as its citizens do not live passively and inertially, taking for granted the social and institutional forms to which they belong, but continually nurture the sense of being of that institution by taking a position regarding its assumptions, aims and objectives, which make of it the state that it is. The more the citizens are subjects of personal acts, the more they contribute to the good life of their state, and the more they individualize and self-constitute themselves in a personal sense. The personal living and acting of citizens as parts of the personal whole-state involves the unfolding of individuals' personality in a public-institutional direction, not in a private one.

As I have already mentioned, Stein shares the same claim: the more citizens act personally, the better they act institutionally (Stein, 1925, II, §§ b–c; see *supra*, § 4.3.1.3). But she develops this perspective further in dealing with it in an entire book devoted to the life of the state-whole as the life of its citizens-parts. Indeed, in her social ontology Stein elaborates important Husserlian social-ontological insights scattered throughout his manuscripts, posthumously collected in *Ideas II* and *Phenomenology of Intersubjectivity*, and in her *Individual and Community* and *An Investigation Concerning the State* she provides a systematic account of them. What is at stake in both Husserl's and Stein's personally embedded social ontology is always the crucial issue of the relationship between the personal and the social being. Husserl's concept of wholes as "unitarian foundation" (Husserl, 1901) provides a fruitful key also adopted by Stein, as we have seen, for understanding the double and mutual constitution of persons and social collectives: from the persons-parts to the collective-whole founded on its parts, and from the collective-whole to the persons-parts that are individualized from the whole to which they belong. This is a mutually existential dependence relationship that needs to be specified and clarified each time, according to the different types of collectives and the personal identities of the individuals involved. How does the fact of belonging to a certain social whole contribute towards making me into the person who I am? And conversely: how does the person I am contribute to the constitution, modification, renovation or decline of the forms of societies and communities to which I belong?[24]

[24] This is a position that characterizes the phenomenologically embedded social ontology and that shows all the problematic nature of what I call the "impermeability thesis" of mainstream social ontology, i.e., that which is rooted in the analytic philosophy tradition, namely the idea that the individual's being is impermeable with respect to the collectives to which they may belong. I

4.3.2.4 Personal and Social Renewal

Husserl deals with the idea of this mutually existential relationship of the personal and of the social being in his *Kaizo* papers (Husserl, 1923–1924), where he specifically focuses on the issue of renewal precisely in the perspective of this mutual implication. Indeed, Husserl claims there that there cannot be personal renewal unless there is a social one, and vice versa, and discusses the idea of a social living that is also a personal living in which "an ethical humanity" is realized.

> The community is a personal, so to speak many-headed, and yet united subjectivity. Its individual persons are its "parts", functionally interwoven with each other through multifarious "social acts" (I-you acts, commands, agreements, acts of love, etc.) which spiritually unite person with person. At times, a community functions as many-headed, and yet in a higher sense, "headless": namely, without concentrating itself into the unity of willing subjectivity and without behaving analogously to a single subject. However, the community can also take on these higher forms of life and become a "personality of higher order". As such, it accomplishes communal achievements which are not merely aggregates of the achievements of individuals, but rather are in a true sense personal achievements of the community as such, realized in its striving and willing. Accordingly, the active life of a community, of an entire people is also able […] to take on the uniform shape of practical reason, the form of an "ethical" life. (Husserl, 1923–1924, En. tr. 2022, *Renewal as Individual-Ethical Problem*: § 1)

A community is a "higher-order personality" in the proper sense if it acts as a subject of a unitarian will. Higher-order personalities are genuine collective subjects because they are unitary wholes irreducible to the sum of their parts-persons: there is an overall will of the personal whole that is not simply the sum of the individual wills of the members of the whole. There is, in other words, the constitution of a *we* as the agent of practical collective intentionality that is irreducible to the mere sum of an *I, you, she*—each with its own individual intentionality. The ethical life of a community is one in which the wills of the persons-parts of the community-whole constitute a unified will.

But this also means that the unitary will of the collective subject is that of a personal whole, agent and author of its own practical life, which in turn retroacts on the lives of individual person-parts by individualizing them in a personal sense. While the individual wills of the person-parts give rise to the unitary will of the collective-whole, at the same time the latter directly intervenes and impacts on the lives of the person-parts by deepening and re-empowering their personal being. This is, again, the twofold direction of foundation—bottom-up, from the parts to the whole, and top-down, from the whole to the parts—that characterizes social wholes as emergent wholes (as we have already seen in Stein's analysis of personal communities; see *supra* §§ 4.3.1.1 and 4.3.1.2), and that in the case of personal wholes such as the ethical life of a community is affirmed in the mode of a foundation that has a personalizing efficacy.

---

developed this point in De Vecchi (2022, Ch. IV). On this issue, see also Zahavi (2021) and above all Zahavi (2025).

According to Husserl, individual ethics always implies social ethics: "ethics is not merely individual ethics, but also *social ethics* [...], there is also necessarily an ethics of communities as communities" (Husserl, 1923–1924, En. tr. 2022, *Renewal as Individual-Ethical Problem*: § 1). Every personal life can be ethical only as part of personal wholes, and this is to be understood precisely in the double direction of foundation of the personal wholes as emergent wholes, as we have seen.

Husserl further emphasizes that ethical life, both individual and social, is a life that does not rest and shrink into an "already given", which comes to be taken for granted, and that does not forget the living experience that founded it. On the contrary, "the ethical life is, according to its essence, a life of a *renewal*, a life from an originary, and then always able to be reactivated will to renewal" (Husserl, 1923–1924, En. tr. 2022, *Renewal as Individual-Ethical Problem*: § *Conclusion*), a will to renewal that cannot be solely individual but must concern the whole community to which the personal individual belongs. Indeed, the problem of renewal is both an "individual ethical and social ethical problem" (Husserl, 1923–1924, En. tr. 2022, *The method of the investigation into essence*).

But what exactly does "renewal" mean as an individual and social ethical problem? It means that all personal life in a genuine sense can unfold only in a constant position-taking attitude on what we are given to live: "*alles Leben ist stellungnehmen*", "all life is position-taking", declares what may be considered one of the adages of Husserlian phenomenology (Husserl, 1911: 56, En. tr.: 290). A necessary condition for ethical living is a constant personal response to the given-data of one's own experience, so as not to reduce one's life, passively, to the flow of one's own vital and psycho-physical tendencies, nor to leave it at the mercy of external pressures and influences.

4.3.2.5 Dynamicity of the Social Landscape

What, then, is the relation between the personally living of position-taking and the living of already given and acquired answers, of traditions and the passive and inertial mode of being of "*de facto* communities"? Husserl also investigates the underlying, obscure and swampy level, which is the terrain proper to the involuntary, the unaware, the comfortable lingering in the layers of personal being in which lurks that which inertially and anonymously goes unchallenged, and which aggregates and unites individuals to one another in an impersonal way. The idea suggested by Husserl is that "higher-order personalities" also presuppose a common fabric of experience that is, so to speak, pre- or sub-personal: a basic horizon of co-belonging, such as that constituted by the simple fact of speaking the same mother tongue, in which people live together in forms of social wholes that are not yet personal wholes. This then means that the common surrounding world, the personal world, the world of personally shared values and meanings as correlates of personal acts (and thus not of values and meanings merely absorbed or imposed by others) is in continuous relation to a more anonymous and impersonal background, which is also a common world of conformity, homologation, imitation. With respect to it, individuals can take a position—i.e., activate themselves as subjects of personal acts, questioning the foundations of that world, questioning it as a common world. In this case, individuals are no longer merely parts of social wholes in which they are connected to each other by passive and unaware experiences that feed a basic social terrain in which they are implicitly ensconced and in which they tend to become inertially entangled. On the contrary, they also become parts of personal wholes in which they are bound to one another by virtue of personal position-takings, and the personal wholes constituted by them in turn exert on the persons-parts an individualizing action that continually fosters and demands their personal unfolding.

Husserl's social ontology is thus a dynamic ontology that has at its core the relationship of continuity and discontinuity between social wholes that are also "personal wholes" and social wholes that are not. This duplexity is, again, well captured in the density of the expression *Gemeingeist*, which indeed occurs in and titles some of Husserl's manuscripts which were included in *Ideas II* and the *Phenomenology of Intersubjectivity*.[25] It is from this standpoint that, at the beginning of my paper, I spoke of an Husserlian gaze attempting to capture the dynamic and changing sequence of scenes of the social landscape, precisely because this social landscape is constantly shifting and evolving, while Stein's gaze zooms in on a freeze-frame of the social landscape, and by patiently chipping away at it she digs up meticulously all the parts of these scenes. They do two different jobs that are equally valuable for a personally embedded qualitative social ontology.

References

Caminada, E. (2019). *Vom Gemeingeist zum Habitus: Husserls Ideen II. Sozialphilosophische Implikationen der Phänomenologie*. Springer.

Caminada, E. (2022). Things, goods, and values: The operative function of Husserl's unitary foundation in Scheler's axiology. *Phenomenology and Mind, 23*, 80–142.

Cimino, A., Moran, D. Staiti, A. (Eds.). (2022). Phenomenology, axiology and metaethics. *Phenomenology and Mind, 23*.

Conni, C. (2005). *Identità e strutture emergenti. Una prospettiva ontologica della Terza ricerca logica di Husserl*. Bompiani.

De Monticelli, R. (2019). Perceiving values—A phenomenological approach. In M. Mühling (Ed.), *Perceiving truth and values* (pp. 43–62). Vandenhoeck & Ruprecht.

De Monticelli, R. (2021). *Towards a phenomenological axiology. Discovering what matters*. Palgrave Macmillan.

De Santis, D. (2025). "Etwas Neues" or what is means to be an "individual idea". *Rivista di Estetica, LXV*(89), 120–135.

De Vecchi, F. (2014). Three types of heterotropic intentionality. A taxonomy in social ontology. In A. Konzelman Ziv & H. B. Schmid (Eds.), *Institutions, emotions and group agents. Contribution to social ontology* (pp. 117–137). Springer.

De Vecchi, F. (2015). Edith Stein's social ontology of the state, the law and social acts. An eidetic approach. *Studia Phaenomenologica, XV*, 303–330.

[25] The expression *Gemeingeist* occurs in *Ideas II*, both in the old and new edition (Husserl, 1912–1928/1952, § 51; 1908–1930/2025: 196, 146). Concerning the role of the *Gemeingeist* in Husserl phenomenology, both in *Ideas* and in *Phenomenology of Intersubjectivity* manuscripts, I refer to Caminada's accurate reconstruction (Caminada, 2019).

De Vecchi, F. (2016). The existential quality issue in social ontology: Eidetics and modifications of essential connections. *Humana.mente. Journal of Philosophical Studies, 31*, 187–204.
De Vecchi, F. (2017). Eidetics of law-making acts: Parts, wholes and degrees of existence. *Phenomenology and Mind, 13*, 86–95.
De Vecchi, F. (2019). Eidetics of empathy, intersubjectivity, embodiment and qualitative ontology. Rediscovering Edith Stein's Account of Empathy. *Humama.mente. Journal of Philosophical Studies, 36*, 221–243.
De Vecchi, F. (2020). Common-surrounding world and qualitative social ontology—Phenomenological insights for the environment and its crisis. *Rivista di Estetica, 3*, 33–51.
De Vecchi, F. (2022). *La società in persona. Ontologia sociale qualitative*. Il Mulino.
De Vecchi, F. (2024). Eidetics of social acts vs. eidetics of acts of meaning: From different wholes-parts connections to different kinds of acts. In P. Alves, P. Theodorou, & A. I. Baka (Eds.), *Phenomenology of Law* (pp. 99–131). Springer.
De Vecchi, F. (2025). Wilhelm Schapp's qualitative social ontology: The role of values for the law. In D. Nucilli & D. De Santis (Eds.), *The philosophy of Wilhelm Schapp. From phenomenology to jurisprudence and the hermeneutics of stories* (pp. 143–170). Bloomsbury.
Embree, L. (2003). *The possibility of a constitutive phenomenology of environment*. In C. S. Brown, & T. Toadvine (Eds.), *Eco-phenomenology. Back to the earth itself* (pp. 37–50). State University of New York Press.
Fuchs, T. (2014). The virtual other: Empathy in the age of virtuality. *Journal of Consciousness Studies, 21*(5–6), 152–173.
Gilbert, M. (2013). *Joint commitment. How we make the social world*. Oxford University Press.
Husserl, E. (1901). *Logische Untersuchungen*. In Hua XIX, ed. U. Panzer. Martinus Nijhoff, 1984. English translation by J. N. Findlay, *Logical Investigations I–II*, Routledge, 2001.
Husserl, E. (1905–1920). *Zur Phänomenologie der Intersubjektivität. Texte aus dem Nachlass. Erster Teil 1905–1920*, in Hua XIII, ed. I. Kern Den Haag: Martinus Nijhoff, 1973.
Husserl, E. (1908–1930/2025). *Ideen zu einer reinen Phänomenologie und phänomenologischen Philosophie. Zweites Buch: Phänomenologische Untersuchungen zur Konstitution und Wissenschaftstheorie*. In Hua IV/V, D. Fonfara (Ed.). Springer.
Husserl, E. (1911). *Die Philosophie als strenge Wissenschaft*. In Hua XXV, *Aufsätze und Vorträge 1911–1921*, eds. Th. Nenon e H. R. Sepp. Martinus Nijhoff, 1987, 2–62. English translation, *Philosophy as a Rigorous Science*, by M. Brainard. *The new yearbook of phenomenology and phenomenological philosophy* 2, Routledge, 259–295.
Husserl, E. (1912–1928/1952). *Ideen zu einer reinen Phänomenologie und phänomenologischen Philosophie. Zweites Buch: Phänomenologische Untersuchungen zur Konstitution*. In Hua IV, ed. M. Biemel. Martinus Nijhoff, 1994. English Translation by R. Rojcewicz and A. Schuwer, *Ideas pertaining to a pure phenomenology and to a phenomenological philosophy. Book 2: Studies in phenomenology of constitution*, Martinus Nijhoff, 1989.
Husserl, E. (1913). *Ideen zu einer reinen Phänomenologie und phänomenologischen Philosophie. Erstes Buch: Allgemeine Einführung in die reine Phänomenologie*. In Hua III/1–2, ed. K. Schuhmann. Martinus Nijhoff, 1950. English translation by. F. Kersten, *Ideas pertaining to a pure phenomenology and to a phenomenological philosophy. First Book. General introduction to a pure phenomenology*, Martinus Nijhoff, 1982.
Husserl, E. 1919. *Natur und Geist. Vorlesungen Sommersemester 1919*. In HuaMat IV, ed. M. Weiler. Kluwer, 2002.
Husserl, E. (1921). *Gemeingeist I: Person, personale Ganze, personale Wirkungsgemeinschaften. Gemeinschaft – Gesellschaft* (pp. 164–184). In E. Husserl 1921–1928/1973b.
Husserl, E. (1918–1921). *Gemeingeist II. Personale Einheiten höherer Ordnung und ihre Wirkungskorrelate* (pp. 192–204). In Husserl 1921–1928/1973b.
Husserl, E. (1921–1928). *Zur Phänomenologie der Intersubjektivität. Texte aus dem Nachlass. Zweiter Teil 1921–1928*. In Husserliana XIV, I. ed. Kern. Martinus Nijhoff, 1973a.
Husserl, E. (1923–1924). *Kaizo-Artikel*. In Husserliana XXVII, *Aufsätze und Vorträge 1922–1937*, eds. Th. Nenon e H.R. Sepp. Den Haag: Kluwer, 1988. English Translation by Buckley, Ph.

Essays on Culture (Kaizo). Unpublished translation draft discussed during the online reading group 2021/2022 of the Husserl Archives, KU Leuven.

Kekki, M-K. (2020). Authentic encountering of others and learning through media-based public discussion: A phenomenological analysis. *Journal of Philosophy of Education, 54*(3), 207–520.

Kohàk, E. (2003). *An understanding heart: Reason, value and transcendental phenomenology*. In C. S. Brown, & T. Toadvine (Eds.), *Eco-phenomenology. Back to the earth itself* (pp. 19–36). State University of New York Press.

Melle, U. (1996). *Nature and spirit*. In T. Nenon & L. Embree (eds.), *Issues in Husserl's Ideas II*, Contributions to Phenomenology (Vol. 24, pp. 15–35). Kluwer.

Moran, D., & Szanto, T. (Eds.). (2016). *Phenomenology of sociality: Discovering the We*. Routledge.

Osler, L. (2024). Taking empathy online. *Inquiry, 67*(1), 302–329.

Reinach, A. (1913). *Die apriorischen Grundlagen des bürgerlichen Rechts*. In A. Reinach, *Sämtliche Werke*, eds. K. Schuhmann and B. Smith. Philosophia Verlag, 141–278, 1989. English translation, *The A priori Foundations of the Civil Law*, by. J. Crosby. In *The A priori Foundations of the Civil Law. Along with the lecture "Concerning Phenomenology"*, ed. J. Crosby with an introduction by A. McIntyre. De Gruyter, 2012, 1–142.

Salice, A., & Schmid, B. (Eds.). (2016). *The phenomenological approach to social reality. History, concepts, problems*. Springer.

Scheler, M. (1913–1923). *Wesen und Formen der Sympathie*. In M. Scheler, *Gesammelte Werke* VII, ed. Maria Scheler. Francke, 1973. English Translation: *The nature of sympathy* by. P. Heath. Routledge & Keegan Paul, 1973.

Scheler, M. (1913–1926). *Der Formalismus in der Ethik und die materiale Wertethik. Neuer Versuch der Grundlegung eines ethischen Personalismus*. In M. Scheler, *Gesammelte Werke* II, ed. M. S. Frings. Bouvier Verlag, 2009. English translation by M. S. Frings and R. L. Fung, *Formalism in Ethics and Non-Formal Ethics of Values*. Northwestern University Press, 1973.

Searle, J. R. (1995). *The construction of social reality*. Routledge.

Searle, J. R. (2010). *Making the social world. The structure of human civilization*. Oxford University Press.

Stein, E. (1917). *Zum Problem der Einfühlung*, ESGA V. ed. M. A. Sondermann. Herder, 2008; English translation: *On the Problem of Empathy* by W. Stein. Springer, 1964.

Stein, E. (1922). Beiträge zur philosophischen Begründung der Psychologie und der Geisteswissenschaften. In E. Stein, *Beiträge zur philosophischen Begründung der Psychologie und der Geisteswissenschaften – Eine Untersuchung über den Staat*. Niemeyer, 1970, 2–284. English translation. *Philosophy of Psychology and the Humanities*, by M. C. Baseheart and M. Sawicki, CWES 7, ICS Publications, 2000.

Stein, E. (1925). Eine Untersuchung über den Staat. *JPPF* 7, 1–123. ESGA VII, ed. I. Riedel-Spangenberger, Herder, 2006. English translation: *An Investigation Concerning the State*, trans. M. Sawicki, CWES 10, ICS Publications, 2006.

Taieb, H. (2020). Acts of the state and representation in Edith Stein. *Journal of Social Ontology, 6*(1), 21–45.

Tönnies, F. (1887). *Gemeinschaft und Gesellschaft: Grundbegriffe der reinen Soziologie*. Wissenschaftliche Buchgesellschaft, 1991.

Vendrell Ferran, I. (2022). Value-feeling and emotional response: Origins and strengths of the alternative to the perceptual model. In C. Erhard & B. Vassilicos (Eds.), *New yearbook for phenomenology and phenomenological philosophy*. Reinach and Contemporary Philosophy (Vol. 2, 151–73).

von Hildebrand, D. (1922). Sittlichkeit und ethische Werterkenntnis. Eine Untersuchung über ethische Strukturprobleme. *JPPF, V*, 462–547.

Zahavi, D. (2014). *Self and other: Exploring subjectivity, empathy, and shame*. Oxford University Press.

Zahavi, D. (2021). We in me or me in we? Collective intentionality and selfhood. *Journal of Social Ontology, 7*(1), 1–20.

Zahavi, D. (2025). *Being we. Phenomenological contributions to social ontology*. Oxford University Press.

Francesca De Vecchi is Full Professor of Theoretical Philosophy at San Raffaele University, Milan. Her research focuses on social ontology, phenomenology of the person, empathy, social acts, and phenomenology of gendered personal identity. She works on a phenomenologically embedded social ontology called *qualitative social ontology*. Her last book is *La società in persona. Ontologia sociale qualitativa* (Il Mulino, 2022). She is the editor in chief of *Phenomenology and Mind* and the director of the *Research Centre in Phenomenology and Sciences of the Person* (PERSONA) at San Raffaele University.

Chapter 5
Us & Them: Edith Stein/T.W. Adorno and Socially Constituted Identity

Pól O'Hagan

Abstract According to T. W. Adorno, any understanding of his philosophical programme hinges on a critical appreciation of his foundational works, "Against Epistemology: A Metacritique", and "Negative Dialectics". Both advance a developed criticism of Husserl's phenomenology and its antecedent philosophical underpinnings. However, some key tenets of Adorno's critique bear a striking similarity to positions advanced by Edith Stein over two decades earlier. Significantly, Stein's understanding of intersubjectivity addresses the entrenched individualism which Adorno found so problematic in the work of Husserl and Heidegger.

Re-engagement with Stein through Adorno (and Adorno through Stein) opens the potential for a dialogue which cogently urges a re-appraisal of the philosophical discourse within Europe, particularly during the inter-war period, calling attention to the systemic centralisation of narratives which have rendered absent, or designated as peripheral, key contributions. Stein's work offers a series of refinements to Adornoian Philosophy, which consequentially represent an opportunity for the development of a new axis of critical thinking in terms of a "Phenomenology of Creativity". This challenges some received orthodoxies, entrenched dogmas and inherited histories which have curtailed and restricted discursive practice across the field of study in relation to Culture and Identity, and heralds an innovative re-alignment of Frankfurt School Critical Theory in phenomenological terms.

Keywords Adorno · Critical Theory · Cultural Studies · Communitarianism · Empathy · Existentialism · Frankfurt School · Heidegger · Husserl · Identity Theory · Intersubjectivity

P. O'Hagan (✉)
Independent Scholar, Belfast, UK
e-mail: p.o-hagan@ulster.ac.uk

M. Lebech (ed.), *Stein and Husserl's Phenomenology of the Social in Ideas II*, Women in the History of Philosophy and Sciences 25,
https://doi.org/10.1007/978-3-032-17339-3_5

5.1 Prologue

"Us and Them, And After All We're Just Ordinary Men,
..." (Waters & Wright, 1973).

[Cut:] Drumheads fill to a faded organ, which oscillates with overtones of a chord held consistently in suspense amidst spilled taped conversation.

[Edit:] Déjà-entendu piano hung upon its minor major seventh chord. Slow modulation—a rising crescendo over a falling Precision bass.

[Play:] An old photograph lyrically fades in; crowded people, attending a protest demonstration, some in military dress, some in civilian suits, hats and attire that leave the impression that they have come together for a constitutional afternoon stroll. Wealthy, well-dressed women look expectantly at the central figure, a man too intense for words, a prophet of the age of ruin; none will meet his gaze.[1]

[Dissolve:] Delayed vowels crashing sibilant into the hiss of a syllabic monad—incomplete and hymnal, its broken harmonies duel mordantly around the dominant substitute in a perfect modal denouement that fails to resolve. Pensive/contemplative looking directly at the viewer—a woman uncertainly foregrounded in her family photograph.[2]

5.2 Introduction

This chapter argues that the philosophical contributions of Edith Stein and Theodor W. Adorno offer complementary yet distinct frameworks for understanding identity as a socially mediated and historically situated process—one that resists reduction to fixed categories such as assimilation, authenticity, or apostasy. By placing Stein and Adorno into critical dialogue with each other, the chapter aims to foreground the *Lebenswelt* as a contested space in which identity is not merely expressed but continually negotiated under conditions of cultural tension and historical transformation.

Both Stein and Adorno engaged deeply with Husserl's phenomenology in their early work,[3] but each moved beyond its methodological confines to confront the implications of identity formation under the pressures of modern social life. Their shared experience as Jewish intellectuals in interwar Germany—navigating a society that simultaneously demanded integration and marked them as other—provided a

---

[1] Reference to a picture of Adolf Hitler, Alfred Rosenberg and cohort of supporters in Munich 1923; The History Blog » 2013 » June » 14 accessed 31/12/21. https://www.gettyimages.co.uk/detail/news-photo/adolf-hitler-alfred-rosenberg-and-dr-friedrich-weber-of-the-news-photo/2637109.

[2] Reference to photograph of Edith Stein (https://en.wikipedia.org/wiki/File:Edith_Stein-Student_at_Breslau_(1913-1914).jpg).

[3] Stein E. (1916), On the Problem of Empathy; Adorno T.W. (1924), The Transcendence of the Linear and Noematic in Husserl's Phenomenology. Both of these supervised PHD dissertations were written for assessment; Stein's supervisor was Edmund Husserl, while Adorno's was Hans Cornelius.

lived context in which philosophical reflection on subjectivity became inseparable from political and cultural realities. Their work thus exemplifies the intersection of phenomenology and cultural critique, articulating a mode of thought in which identity is neither a stable possession, nor a passive inheritance, but a dynamic, contested construction shaped by structures of power, exclusion, and remembrance.

Despite this, assessments of both thinkers have often been coloured by retrospective moralisation: Stein's conversion to Catholicism has been read as a renunciation of her Jewish heritage, while Adorno's critical stance toward mass culture and Enlightenment has been interpreted as alienated pessimism. Such readings, however, obscure the more nuanced and historically embedded nature of their thought, which resists simple binaries and instead insists on the dialectical tension between individual agency and social determination.

As Shmuel Feiner (2004) has demonstrated in his study of the *Haskalah*, terms such as "assimilation" function reductively when applied as blanket descriptors of Jewish modernity. They risk collapsing the complexity of lived experience into linear narratives of loss or betrayal.[3] By extension, they also misrepresent the critical positions taken by figures like Stein and Adorno, whose philosophical work reflects not capitulation to dominant norms, but an acute awareness of the compromises and contradictions imposed by them.[4]

By attending to the tensions embedded in the *Lebenswelt*, this chapter contends that Stein and Adorno offer vital resources for rethinking identity beyond polarised debates. Their work illuminates how identity is constituted through historically embedded negotiations of difference, and how philosophical reflection might critically engage with these negotiations without reducing them to moral judgments. In doing so, they open up a cultural-theoretical space for understanding subjectivity as a site of both vulnerability and resistance within the modern social order.[5]

5.3 Background

Although the Prussian Emancipation Edict of 1812 effectively gave the Jewish community civil rights, privileges, and protections, allowing land ownership and the right to take up municipal or university posts, prohibitions remained on officer entitlement in respect of military service, and there continued to be restrictions on legal and governmental roles. The city of Breslau became an attractive destination for Jewish families: the small community (population 3522 in 1810) increased decade upon decade, as it grew to become the sixth largest city in the German Empire. After unification in 1871, Breslau was known as a stronghold of left-wing liberalism with

[4] Stein's *On the Problem of Empathy* (1917/2016) and Adorno's early writings on phenomenology demonstrate parallel attempts to reckon with intersubjectivity under social constraint.

[5] On Stein's ambivalent relation to Jewish custom and her later theological development, see Sawicki (1997); on Adorno's navigation of Jewish identity within critical theory, see Müller-Doohm (2005).

a diverse population, which tripled to over half a million in the period between 1860 and 1910 (Brilling, 2008).

Post WWI, Breslau was designated the capital of a newly created Prussian province of Lower Silesia, which was part of the Weimar Republic. Settlement had been galvanized by imperial and subsequently municipal protection, which acknowledged the important business contribution of the Jewish community and established a system of toleration where categories of *Schutzjuden* (protected Jews) were able to establish communities, albeit in designated areas of the city. Most of the prominent Jewish families in Breslau favoured *Haskalah* or reformist tendencies and established schools promoting a secular curriculum and education for girls (something that was opposed by the orthodox rabbinical community which also had a strong presence in the city). As the youngest of eleven siblings in a Jewish family living in Breslau, Stein would have been intimately aware not only of innate tensions within her own community, but also of the contested status of the city itself.

This erupted throughout the 1920s, with a number of violent Polish Silesian insurrections which attempted to instigate a secession from the newly established Weimar Republic. Reprisals in August 1920 by the German Freikorps—an irregular German paramilitary organisation formed to protect the German Reich, after its post WWI defeat—led to the destruction of the Polish Consulate, and the burning down of the Polish Library by a mob. These pogroms resulted in the number of Poles in the city falling significantly. The Freikorps, implicated in summary executions, assassinations and violent beatings, were tolerated by the German state, who gave them semi-legitimacy, using them to quell socialist revolts, break up workers' strikes, put down ethnic or nationalist uprisings in Silesia and the Baltic States and to generally suppress civil unrest (Caballero Jurado & Bujeiro, 2001). While in 1925, the Jewish population in Breslau had risen to over 23,000, despite anti-Semitic riots in 1923, in stark contrast, by 1933, when the NSDAP acceded to power, Breslau had become one of the party's strongest support bases. In the 1932 elections the 'Nazi's' received 44% of the city's vote, their third-highest total in all of Germany (Van Rahden, 2008, pp. 323–6).

In view of the historical significance of events in her hometown, Stein's conspicuously light, anecdotal "memoire" *Life in a Jewish Family* (2016) offers a layered textual subtlety that only becomes apparent once the reason for her writing is understood. Her foreword clearly states her motivation:

> "Recent months have catapulted the German Jews out of the peaceful existence they had come to take for granted. They have been forced to reflect upon themselves, upon their being, and their destiny. But today's events have also impelled many others, hitherto non-partisan, to take up the Jewish question [...] a Jewish friend of mine expressed her anguish: "If only I knew how Hitler came by his terrible hatred of 'the Jews' [...]" From these sources, as though from a concave mirror, a horrendous caricature looked out at us." (Stein, 2016, p. 25).

Catholic Resistance Literature [CRL], produced during the 1930s to undermine the Nazis' anti-clerical, antisemitic propaganda campaign (which was officially endorsed by the Church in programmed writings and speeches), attacked the pseudo-religious racial theories of Nazi ideologues such as Alfred Rosenberg. Exemplified by Bishop Clemens Von Galen's vehement castigation, "The so-called eternal racial soul [...]

is in reality a nullity." (Von Galen in Evans 2002, p. 234), CRL pamphlets and statements were disseminated widely through church circulars and devotional groups, as well as across university campuses.

Edith Stein's position as a teacher in a Catholic education college in Münster placed her directly in the front line of Rosenberg's ideological education policies and its racist agenda. In April 1933, Stein wrote to Pope Pius XI, expressing moral outrage at the Nazi regime's treatment of Jews: "For weeks, we have seen deeds perpetrated... [which] originate with a government that calls itself 'Christian.' [...] We are convinced that this silence will not, in the long run, purchase peace with the present German government." (Stein, 1933).

Given this context, Stein's account of her Jewish upbringing can be considered a calculated piece of populist writing designed to counter the caricatures, misrepresentations and tropes propagated by Nazi controlled government agencies. Significantly Sawicki observes that the "role Stein may have played in the Münster challenge to Rosenberg is a question worthy of further research" (Sawicki, 1997, p. 170).

Written for a conservative German Catholic audience, Stein's autobiographical account can itself be interpreted as a carefully constructed act of resistance and defiance aimed at those who still might have misgivings about supporting the new policies of racial discrimination. Her presentation of an "assimilationist" Jewish community resonates with conservative Catholic values and directly challenges Nazi appropriation of the term *Gemeinschaft*, which emphasised ethnic homogeneity as the foundation of a distinctly German national identity. Stein's profound philosophical understanding of how identity is socially constructed imbues each anecdote about family life with a resonant subtext that cannot be readily elicited without an understanding of the socio-political events that inspired its inception in 1933. In view of this, it could be concluded that Stein's "memoire" purposefully attempts to author and construct a social identity for political reasons, albeit inconspicuously—it should be noted that Stein gave instructions that these writings remain unpublished until after the death of her last surviving family member (Stein, 2016, p. 6).

In many respects, Theodore Wiesengrund-Adorno's family background contrasts distinctly with that presented in Stein's autobiography. As the only male child of a mixed marriage between an affluent secular Jewish wine importer and a Catholic Corsican opera singer, Adorno enjoyed the privileges of social and financial capital. His home in the cosmopolitan, ethnically diverse and politically progressive city of Frankfurt Am Main, itself a major centre of the *Haskalah* reform movement, supported his precocious academic talent. Interestingly, Adorno's late adoption of his mother's name (Von Boeckmann, 2004, pp. 62–64) was the basis, in 1966, for a castigating aspersion by Hannah Arendt. Writing to Karl Jaspers, Arendt accused Adorno of being complicit with the Nazi regime, in attempting to pass under dispensations allowed for *Mischlinge* (partial Jews) under the racist Nuremburg laws of 1935:

> Adorno's unsuccessful attempt at Gleichschaltung had been revealed by the Frankfurt student paper Diskus. He replied with an indescribably pathetic letter which has none the less made a great impression on the Germans. The real infamy was that he a half-Jew among genuine

> Jews took this step without informing his friends. He hoped to get through with his mother's Italian side, Adorno versus Wiesengrund. (Arendt quoted in Wilcock, 2000, p. 174).

Evelyn Wilcock, adopting a strong defence of Adorno, makes it clear that Arendt is demonstrably wrong. There is no doubt that Adorno was a victim of the Nazi regime, having lost his university position and been forced into exile as a result of the Nuremburg Laws. Additionally, Gillian Rose points out that "in September 1933, on his thirtieth birthday, [Adorno] was deprived of his *venia legendi* (right to teach) by the Nazis [...] In the spring of 1934 he left Germany and came to London" (Rose, 2014, p. 9). Adorno's relationship with Gretel Karplus, who was also a Jewish intellectual and a capable businesswoman, as well as his association with Marxist Jewish intellectuals, marked him out as a radical and undoubtedly put him, as well as his family and associates in real danger. Wilcock points out that under the racist Nuremburg Laws it was forbidden for *Mischlinge* to marry a spouse from a "full" Jewish background (i.e., where both parents were Jewish). Effectively this meant that, given their status, it was illegal for Adorno and Gretel Karplus to wed under Nazi legislation (Wilcock, 2000, pp. 175–177). Arendt's racial slur and her assertion of ethnic authenticity, play into a hierarchy of victimhood which obfuscates the contradictions inherent in the constitution of identity and wilfully ignores the negotiation of social constrictions necessarily imposed by the pejorative system of racial profiling. It is therefore entirely obvious that Arendt's *ad hominem* attacks on Adorno, were motivated by reasons beyond a criticism of his work, and were designed to damage and impugn his professional reputation (Wilcock, 2000).

From an analytical perspective, both Stein and Adorno are clearly concerned with constructing a critical defence against the justifications for discrimination, subjugation, and brutality, manifested in the pseudo-science of racial categorization and the systemization of identity—itself a hangover from nineteenth century imperialist nation building. Both theorists attacked reductive notions of authentic identity and were forthright in their criticism of Martin Heidegger, whose own nationalistic sympathies[6] (curiously defended by Arendt) are evidenced in his own unapologetic participation as a member of the NSDAP (Wilcock, 2000). Elisabeth Young-Bruehl has, moreover, drawn attention to the fact that Hannah Arendt while she was a student of Heidegger, became his lover, a fact little known until after both of their deaths (Young-Bruehl, 2004).

In acknowledging that the "experience of controversial social identity [...] marked Stein's thought significantly" (Lebech, 2015, p. 7), it is important to also state that for Stein, identity is necessarily relational, dynamic, and ultimately dependent upon the interaction between individuals and communities. Exemplified in her life, work, and ultimately her death in Auschwitz, the ramifications of socially constituted identity

[6] Martin Heidegger joined the National Socialist German Workers Party on first May 1933 (Farin, Ingo 2016, Reading Heidegger's "Black notebooks" 1931–1941. Cambridge, Massachusetts London, England: The MIT Press. ISBN 978–0262034012). (Sharpe, Matthew 2 October 2018, "On Reading Heidegger—After the 'Heidegger Case'?". Critical Horizons. 19 (4): 334–360. doi:10.1080/14409917.2018.1520514.ISSN 1440-9917. S2CID 149,832,726), (di Cesare, Donatella 2018-Heidegger and the Jews: the Black notebooks. Cambridge, UK Medford, MA: Polity. ISBN 9781509:503834)

continue to echo in controversies surrounding Stein's beatification and her symbolic representation as both apostate and saint. Stein's execution for being Jewish despite her conversion to Catholicism inexorably underlines the conditionality of identity, which operates not just as a case of "who we think we are", but also is determined by what others decide and think about us and what we represent for them.

Adorno observes, that it is in contradiction and paradox that we come to an appreciation of the "sutured" reality of an existence forged out of ambiguity and uncertainty. By contrast, reductive perspectives fail to consider the trace elements of historical experience and neglect the complex web of influence which characterize public, private, internal, and external dialogue; general assumptions about identity then become problematic on multiple counts. Both Stein and Adorno appreciate socio-political and cultural identities as multi-faceted affairs, that become further complicated by the situated reality of familial and work relationships and the impact of historical events. Both theorists forge a critical framework that is distinct from, but indebted to, Husserl's phenomenology, to address what are undoubtedly important philosophical, psychological, and social questions arising from the social constitution of identity, whereby "non-conceptuality, inalienable from the concept, disavows the concept's being-in-itself" (Groff, 2014, p. 168).

5.4 Sonorities & Antinomies

Adorno, galvanized by a sense that the potential of phenomenology had not yet been realised, (Buck-Mors, 1977, Rose, 2014), shares common ground with aspects of Stein's philosophy, notably in the desire to establish a distinct "science of the humanities" based upon appropriate methodologies and practices influenced by the ideas of William Dilthey (Jay, 1973, p. 49; Stein, 2000, p. 370). Moreover, Adorno's *Negative Dialectics* (2007) as an "anti-system" philosophy, purposefully linked to his earlier work *Against Epistemology* (2013), and a direct criticism of Husserl's early emphasis on individualism, hinges on the same premise which underpins Stein's pursuit of empathy as a problem relating to intersubjectivity. While Stein, however, managed to maintain and advance a position within the bounds of an Husserlian discourse, Adorno's incisive critique indicated a desire to break with phenomenology—a position which belied his obvious and prolonged interest in its role as a fecund methodology. This position was due in large part to the influence of his mentor Hans Cornelius who furnished both Adorno and his mentor/colleague Max Horkheimer with a politicised phenomenological perspective bearing an anti-establishment ethos, which could be productively aligned with Marxism (Jay, 1973, pp. 44–45). Interestingly, Stein's later critique of Heidegger's "philosophy of a bad conscience" (Ballard, 2007; Wargo, 2011; Wilk, 2007) is no less incisive and robust in its rebuttal of individualism than Adorno's caustic reproach, encapsulated by his later polemic attacking Heidegger, '*The Jargon of Authenticity*' (1973).

Marianne Sawicki, an important and prominent expositor of Stein's work, who is recognised as having a critical perspective which engages with the philosophical and societal dimensions of her phenomenological approach, demonstrates that Stein's critical distance from Husserlian orthodoxy differs significantly enough to be considered a hybrid variant, while at the same time remaining in accordance with phenomenology's foundational precepts (Sawicki, 1997). Hence, Stein's emphasis on the empathetic connection between subjects (intersubjectivity) as a constituting factor of individual personhood allowed Stein to develop a philosophical framework which "unavoidably partakes in the dangerous negotiation of difference and correctness" (Sawicki, 1997, p. 141). This is a position that is both non-totalising and inclusive, and one which accords with Adorno's conclusion that the parameters of individual experience cannot be construed as foundational in terms of societal engagement (Adorno, 2007). Stein's recognition of the subject as an entity socially constituted through its relationships with "*alien*" subjects characterised the empathetic relationship as an opportunity "to perceive the alien centre", not in terms of conquest or subjugation, but as a foundation upon which "to agree to a world with multiple centres" (Sawicki, 1997, p. 126). Sawicki observes that Stein's reworking of both transcendental and realist schools of phenomenology created a distinct hybrid where her "phenomenology of understanding is complemented by her phenomenology of creativity" (Sawicki, 1997, p 144).

Although Adorno's Constellation Method attempts in many respects to build upon his perceived failure of Husserl's programme, his own analysis of Husserl's phenomenology is in some respects significantly flawed. His dismissal of the work of Husserl's "disciples" in his "metacritique" (2013) fails to take account of important theoretical developments advanced by Stein in dealing with problematic aspects of Husserl's philosophy. Significantly, Adorno neglected Husserl's *Ideen II*, a work delayed in publication, which contained changes and qualifications to Husserl implemented by Stein (Moran, 2017; Nenon & Embree, 2013), rectifying much of the contentious material upon which Adorno based his earlier criticism of Husserl's transcendental idealism. Importantly, Stein's prioritisation of "the original current of consciousness" as "a pure becoming" (Stein, 2000, p. 23) distinctly posits a materialist fluctuating experiential state of existence bounded by the corporeality of the body:

> As we consider expressions to be proceeding from experiences, we have the spirit here simultaneously reaching into the physical world "becoming visible" in the living body. This is made possible by the psychic reality of acts as experiences of a psycho-physical individual, and it involves an effect on physical nature (Stein, 1989, p. 88).

Stein's conception of potentiality, expressed in chapter two of her major work *Finite and Eternal Being* (2002), also anticipates Adorno's peer Ernst Bloch by positing the *not-yet* and the *no-longer* as (in)formants which condition the experiential moment of actualised being whereby "the unfolding of the essence is a temporal process" (Stein, 2002, p. 162). For Adorno, essentialist assertions where "a sort of being *sui generis* is passed off" (2013, p. 71) become redundant given that "[n]on-conceptuality, inalienable from the concept, disavows the concepts being-in-itself" (Groff, 2014, p. 168).

Both theorists therefore share a sense of the impermanent, transient nature of existence and rail against the imposition of a static concept of identity or personhood. As Stein eloquently observes: "What does not 'express' itself in vital action remains hidden, not only veiled to our knowledge, but ontically [*seinsmässig*] undisclosed like the inner life of a closed bud" (2002, p. 160). While Adorno castigates the "fraudulent plausibility" of what is "given" *in actio*, and insists upon a pluralistic approach, determining that "one can speak meaningfully (*sinnvoll*) about identity only in relation to multiplicity" (2013, p. 98), Stein ruminates:

> That an experiential content coalesces out of multiple components is [...] nothing new. Indeed, an individual experience too is not something instantaneous. Rather, it develops in a continuity of experience during a certain period of time and shows all sorts of qualitative fluctuations within its unity. (Stein, 2000, p. 136).

Fundamentally for Adorno, the essential condition for any erudite assertion of identity must be an acknowledgement of non-identity, and with it a comprehension of complex causality which means rejecting the concept of authentic being as anathema. This led him to be heavily critical of hermeneutic phenomenology and *Dasein* which, as Paul Ricœur recognised, Heidegger grounded in the act of naming (*nennen*): "the forming of a name [establishes] in its Being a being that is opening itself up and preserves (*bewahrt*) what it is in this open-ness, constriction and constance" (Heidegger, quoted in Ricœur, 2004, p. 228). In privileging this concept of naming, Heidegger raises identity to a mythopoetic sense of authenticity, the seeking out of which existentially becomes *raison d'être* for humanity; "thus naming designates the place and role of man in language" and by consequence releases the potential for a realisation of the self as being-in-the-world" (Ricœur, 2004, p. 228).

For Adorno, the imposition of identity is an attempt to determine relations between subject and object which, he argues, are necessarily those of dominance and subjugation, or ownership and alienation, inscribing an historical moment consecrated by the transactional exchange and abstraction of labour. Drawing attention to the duplicity inherent in the act of exchange, Adorno and Horkheimer provide an exposition of the contractual nature of naming and relate this to conditions of debt and retribution in their critical chapter on Homer's Odyssey in *Dialectic of Enlightenment* (2002). Anonymity, or the use of a false name/pseudonym/alias, which hides the identity of the agent, acts by contrast as the veil of subterfuge, protecting and delaying transactional consequences, the hubris of declaration carrying with it the weight of recompense and retribution (Adorno & Horkheimer, 2002).

Hence, memes of precarity accompany the oppressed throughout often brutal histories, while absence and silence are the hallmarks of domination and subjugation where anonymity becomes a cipher—in structuralist terms, a signifier—onto which symbolic representations and imaginative fantasies can be projected (Adorno & Horkheimer, 2002, pp. 54–56). Non-Being is therefore the receptive principle which renders existence always incomplete. Negation of identity is then the excluding condition of actualised being which the act of naming conspicuously conceals. Hence, the naming of the individual becomes the critical contractual bond of exchange whereby property ownership passes from the producer to the holder of capital. For Adorno, constitution of both name and deed is co-effectual in determining not just attribution but also judgement of retribution and/or reward, while anonymity obfuscates identity in a way that refutes the authority of a judgement visited upon the individual. Stein's work, often unattributed, is characterised by Sawicki as being marked by a dialectic of presence and absence. Lack of attribution, particularly marked in terms of Stein's contribution to Husserl's *Ideen II* (and mirrored in her nickname "*Anonyma*", used by her close circle of friends) is indicative of her acute awareness of the problematic nature of being seen in the world. Hence, Stein's characteristic self-effacement can be understood as a negotiation of hostile spaces where symbolic violence and actual brutality await the transgressive acts of self-assertion and social recognition demanded by individuals who fail to know their place and disturb the social codes of status and entitlement. Sawicki reminds us, however, "that anonymity was also an effective strategy of political resistance" (Sawicki, 1997, p. 152), and this must certainly be taken into account when reflecting on the complex character of Stein and her milieux.

Adorno's contention is that the social lie of individualism promoted by the vehicle of mass culture invokes its negative proponent, a shadow self, by way of generalisation, trope, and stereotype. Thus, the excluded other, treated as a homogenous collective and stripped of individuality, is presented as an absence—a *not-we* that abjures conformity to social norms. Mass psychosis, for Adorno, is a process manufactured through the construction of pseudo-individualities which afford recognition of the familiar while offering the semblance of distinction or variation. It accords with delusions of individuality which have in fact been produced and propagated as a charade of being. However keenly aware Stein is of the process of sociation, nevertheless a lack of engagement with Historical Materialism leaves her philosophy wanting in respect of a solid theory of alienation. Marked by an absence of conflict theory, Stein leans towards a consensus-oriented Durkheimian social contract model of society, which from an Adornoian perspective proves woefully inadequate for any radical exposition of the subject/object dichotomy. Referencing Simmel, Stein argues that the experiential continuous current of community does not coincide with historical occurrence, but "consists in discrete events that are collected under a unified idea [...] If you dissect these unities further and further, you finally come to 'atoms' that no longer have any historical sense and out of which the historical no longer can coalesce" (Stein, 2000, p. 118).

Distinguishing between associative and communitarian aspects of social formation and invoking a somewhat Tolstoyan vision of "genuine community" (Stein, 2000, p. 263), Stein rejects socio-economic forces as determinative, although admittedly she acknowledges that they are what forms the societal state. By positing the notion of "pseudo-formations", however, she commits to a perspective that is shared with Adorno, who insists that cultural bonds are just as important as economics in the determination of societal relationships.

Focusing on the relations of production as opposed to the forces of production, Adorno emphasises the relative autonomy of culture as a force for social cohesion and as a potential disrupter of societal norms and values, thus paving the way for a revision of Marxist theory, which reinvigorated the discourse around culture and ideology and became known as "the cultural turn" (Jameson, 1990). Contrastingly, Stein can come across as politically naïve and immature in terms of social theory; speaking of "social standing" in reference to the mores of "polite society", she dances around concepts such as ideology, alienation and "bourgeois elitism" without any direct reference to Marxist philosophy. This inevitably creates problems in her discussion of social roles and types, because without a theory of alienation, statements such as "all social life and all social modes finally refer back to the core of the person, which is beyond the reach of all the influences of reciprocal communication" (Stein, 2000, p. 290) can be construed as individualist and even Heideggerian (which I would contend is in keeping neither with her ethos nor her philosophical argument).

However, in making distinctions between the organic communities and associations, and in rejecting notions of a self-supporting communal consciousness, Stein again unexpectedly aligns with Adorno, infamous for his rejection of proletarian class consciousness as the condition for revolutionary change, a stance that split him decisively from Hegelian Marxists such as Gyorgy Lukacs (Jameson, 1990). Stein insists that "social layering recurs in all communities of a determinate type. It's of the essence of class that classes cut through those communities, uniting into themselves individuals out of various communities: like 'mass' (proletariat), 'community' is international" (Stein, 2000, p. 287) and this can be read productively from a socialist perspective. When this is aligned to Adorno's conviction that the contradictions inherent in social modes of communication determined by a system of social control reveal "precisely a situation of alienation and reified consciousness in which people do not know themselves to be the substance of their own social ground" (Groff, 2014, p. 150), Stein's philosophy begins to offer a complex profundity through which ideology can be examined as a form of social contagion (Stein, 2000, p. 149). Moreover, Stein's prioritisation of the "sense-coherence of a communal life", where an individual's contribution is "eligible for historical treatment only when they play a role in the collective progress of life which historical science is seeking to trace" (2000, p. 150), supports a kind of unorthodox Marxian reading of phenomenology that is both nuanced and non-dogmatic. Adorno's observation that "[b]eyond the magic circle of identitarian philosophy the transcendental subject can be deciphered as a society unaware of itself" (Groff, 2014, p. 171), also echoes Stein's inquiry into "uncomprehending contagion" as a process of mass stimulation (unavoidably a descriptor of ideology). Moreover, Adorno's own theories on propaganda and mass control, which he developed with Horkheimer in his seminal work on "The Culture Industries" (Adorno & Horkheimer, 2002), are prefigured by these Steinian conceptualisations of infection and contagion:

> What this makes possible is a modification of the behaviour of one individual under the influence of another, a conformity in behaviour of a series of individuals who mutually influence one another, and finally an intermeshing of functionalities of different individuals which serves objectively one purpose [...] It belongs to the substantial function of contagion that you know nothing of the factor of contagion, but rather take the adopted conviction to be your original one. The other and his frame of mind isn't an item for the one who succumbs to the contagion. Here, as always, no mental functioning is required for the process of contagion (Stein, 2000, pp. 118–119).

Importantly, Stein emphasises the importance of fantasy as a principle of exploration that takes us beyond individuality and the rationalised self towards a repertoire of sense that is correlative to a communal experience, with the proviso that "as soon as you go on to the intuitive fulfilment of such sense content, a set of merely private single intuitions takes the place of the communal experience" (Stein, 2000, p. 35). The complex of individual desires and the operation of group psychology is unavoidably the domain of psychoanalysis and in particular the work of Freud's 1921 publication "Group Psychology and the Analysis of the Ego" (1991, pp. 91–167), which Stein may have been aware of, and which Adorno certainly embraced as a materialist adjunct to Critical Theory (Jay, 1973, pp. 86–89). Effusively stating that psychology has its place along with the physical sciences and sociology, Stein nonetheless insisted that a distinct philosophy of the humanities is necessary to deal with qualitative aspects of individuality and community, which she argued are causal and determinative whereby "the formation of community, like the possible configuration and reshaping of the personality, is limited by individual distinctiveness" (Stein, 2000, p. 269). Stein's conception of a hybrid approach whereby "a kind of valence or power gradient obtains between reflector and reflected, knower and known" (Sawicki, 1997, p. 103), as a symptom of the subject/object dichotomy, could well have been appropriated by Adorno for his (meta) critique of Husserl (Adorno, 2013), thus forming the basis of a relational phenomenology which Adorno went on to develop as his "Constellation Theory", which espouses a relational approach emphasising the propensity of art and creativity to provoke crises of meaning. This latter, Adorno insists, reveals the fractures and limitations of societal epistemologies and ontologies laid bare as abstract systems of control and domination or ideology. Systemic denial of plurality and studied ignorance of multiplicities of realities are therefore compounded by the tendency of a totalising structure to homogenise and authenticate its constructed intentionality as a natural condition of existence. For Adorno, "the movement of abstraction allows us to get rid of that from which we abstract. It is eliminated from our thought, banished from the realm where thought is home, but not annihilated in itself; the faith in it is magical" (Groff, 2014, p. 167).

Adorno's concern with the particular as an atomistic reference point which invokes universality is shared by Stein, however much he finds problematic the phenomenological positing of this particular as an authentic essence. Non-identity and non-being, for Adorno, render the conceptualisation of essence an impossibility beyond the mediation of a systemic framework; hence, his famous maxim, "the whole is false", decries the assertion of systemic totality as false consciousness (Adorno, 2005, p. 49).

Interestingly, Stein accords this problem a theological status which lies beyond the remit of philosophical inquiry—hence, "our knowledge may find it impossible to determine whether a change of essence occurs or whether one object is 'annihilated' and *another* object comes into being" (2002, p. 219). Adorno concurs that philosophy fails to provide a scheme of inquiry that can facilitate such questions, but counters that this is the result of its failure to transform the world, and of its complicity with the dominant social structure, to which it owes obedience. Notably, Adorno does not totally reject the possibility of a critical social truth coming from a religious or mystical perspective, although he expresses strong concerns about the authoritarian implications of adopting a religious affiliation. Pointing to the "subterranean influence" of religious themes on the materialism of Critical Theory, Martin Jay observes that the Frankfurt School were not dogmatic atheists and that they rejected the notion that religion was "solely false consciousness, because it effectively helped preserve a hope for future justice, which bourgeois atheism denied" (Jay, 1973, p. 56). An alternative society based on socialist principles demanded apprehension beyond a philosophy which reflected socio-political reality, and initiated "the move away from philosophy *within* philosophy itself" (Cole 2014, p. 94). John Bossy's (1982) observation, that the evolution of the terms "society" and "religion" are inextricably linked both philosophically and historically, allows religious mysticism and utopian materialism to be construed as critical attitudes directed towards society's failings. Calling time on systemic totality in normative terms, by way of an apocalyptic (revelatory) reconstitution, these stances are aligned to a notion of temporality which is pluralistically interwoven and interdimensional, in recognition that "the experiences of a community itself can extend over a much greater duration than those of an individual" (Stein, 2000, p. 149).

As much as Hedwig Conrad-Martius' ideas underpin Stein's complex conceptualisation of the interplay of time and space, Adorno engages Walter Benjamin's Angel of History conceptually as a metaphor for a radical realignment of temporality. This is founded upon Benjamin's idea of reclamation, where history's waste products are effectively read as archaic indices of society's mis-construed idea of itself. Adorno's rejection of history as a linear progressive entity in favour of Benjamin's non-linear notion of time related to loss and recovery, and his adoption of Benjamin's 'Concept on the philosophy of history', allows for a transcendental materialism which diffuses the reductive determinism of mortality by instituting culture as a form of reflective memory. The archaic past which "brushes history against the grain", inaugurating the "*Jetztsein*" and "*Jetztzeit*" of redemptive revelation, affords and instigates the act of redemption of a modernity haunted by the failure of philosophy to change the world, manifest in the desire for societal communion with a lost past prised from the cracks of a fractured present. Likewise, Stein incisively notes, "[t]here is henceforth a split or crack [*Bruch*] even in the determinateness of the essences of things, they are still a mirror of divine perfection but the mirror is broken [revealing] the general corruption of all things in the fallen state" (Stein, 2002, p. 241).

Stein also observes that while ideal or essential being remains timeless, "real being unfolds temporally—from potency toward act and within time and space" (2002, p. 331). This "immanent time (*Immanente Zeit*)"[7] allows for the experience of wholeness and completeness intuitively as opposed to through rationalisation (Tullius, 2019). For Adorno, "the splinter in your eye is the best magnifying glass" (2005, p. 49), and this imbues any assertion of systemic totality with a jaundiced hue; hence, systemised notions of time, space, and history, which are subject to the

[7] Although Stein doesn't identify the term "immanent time," her ontology portrays time as an immanent, structural dimension of real being—a lived unfolding from potentiality into actuality; this is supported by commentators such as William Tullius (2015).

control and domination of ruling interests, leave traces of an eradicated past which echo and haunt its epistemological footfalls.

5.5 Conclusion

There have inevitably been many attempts to forge a phenomenology that can engage productively with Marx (Benhabib, 2003, Rensman & Gandesha, 2012, Jay, 1973). However, as Adorno frequently points out, alignment between Critical Theory and Phenomenology cannot simply brush fundamental differences under the carpet, especially when these have evolved out of deep-seated social, political, and historical disparities, as well as those that are in theoretical dispute. In this respect, examination of common shared principles between Stein and Adorno provides the impetus for a more suitable foundation for "disclosing the presence of pre-existing codes and models" and developing an apprehension of "truth as transcoding" (Jameson 1972, p. 216) than approaches which adopt a Heideggerian mode of enquiry which seek an elusive authentic self.

However much Adorno's Negative Dialectics can be seen to integrate (and perhaps even appropriate) Stein's conception of the socially constituted individual, Stein's conceptualization of sentient causality and social contagion in turn offers a phenomenological accommodation of Adornoian theories of ideology—effectively creating a reflexive and complex philosophical model which avoids the Adornoian tendency to turn critically upon itself in what has been described by some critics as Critical Theory's "infinite regress". (Buck-Mors, 1977, Jameson, 1990, Jay, 1973, Rose, 2014). Furthermore, Stein's relational approach to phenomenology offers the possibility of a range of "*elective affinities*" which take account of differences between the disciplines of law, economics, and cultural studies, while at the same time offering a hybrid approach which combines the philosophical, sociological and psychological dimensions of creative practice. Consequently, overlapping tendencies between the relational phenomenology of Edith Stein and the constellation theory of T.W. Adorno, constitute an axis needing further exploration. In particular, this could address inadequate accounts of socially constituted identity which avoid analysis of societal modes of power, domination and authority, where these are organising principles used to define and label the unentitled, the dispossessed and the alienated as "*other*".

References

Adorno T.W. (1973). *The jargon of authenticity* (K. Tarnowski and F. Will, Trans.). Northwestern University Press.

Adorno, T. W. (1994). *Hegel: Three studies*. MIT Press.

Adorno T.W. (2001). *Kant's critique of pure reason* (1959), ed. R. Tiedemann (R. Livingstone, Trans.). Polity Press.
Adorno T.W., Horkheimer M. (2002). *Dialectic of enlightenment: Philosophical fragments*, (Board of Trustees of the Leland Stanford Junior University, Trans.). Stanford University Press.
Adorno T.W. (2005). *Minima Moralia: Reflections on a Damaged Life* (E.F.N. Jephcott, Trans.). Verso.
Adorno T.W. (2007). *Negative Dialectics* (E.B. Ashton, Trans.). The Continuum International Publishing Group
Adorno T.W. (2008) *Lectures on Negative Dialectics: Fragments of a lecture Course 1965/1966*. ed. R. Tiedemann (R. Livingstone, Trans.). Polity Press.
Adorno, T. W. (2013). *Against epistemology: A metacritique*. (W. Domingo, Trans.). Polity Press.
Ballard, B. W. (2007). The difference for philosophy: Edith Stein and Martin Heidegger. *The Journal of Value Inquiry, 41*(1), 95–105.
Benhabib, S. (2003). *The reluctant modernism of Hannah Arendt*. Rowman & Littlefield.
Bossy, J. (1982). Some elementary forms of Durkheim. *Past & Present, 95*, 3–18. http://www.jstor.org/stable/650730
Brilling, B. (2008). *Jewish Virtual Library*.. Breslau, Poland,https://www.jewishvirtuallibrary.org/breslau accessed 02/05/23.
Buck-Mors, S. (1977). *The origin of negative dialectics: Theodore W. Adorno, Walter Benjamin, and The Frankfurt Institute*, The Free Press–Macmillan Publishing
Caballero Jurado, C., & with Bujeiro, R. (2001). *The German freikorps 1918–23*. Osprey Publishing.
Demel, W. and Puschner, U. (1995). *Von der Französischen Revolution bis zum Wiener Kongreß* 1789–1815. Deutsche Geschichte in Quellen und Darstellung 6. Reclam.
Feiner S. (2004). *The jewish enlightenment* (C. Naor, Trans.). University of Pennsylvania Press.
Freud, S. (1991). *Civilization, society and religion: Group psychology*. Civilization and its Discontents and Other Works.
Groff, R. (Ed.). (2014). *Subject and object: Frankfurt school writings on epistemology, ontology, and method*. Bloomsbury.
Lebech, M. (2015). *The philosophy of Edith Stein: From phenomenology to Metaphysics*. Peter Lang.
Jameson, F. (1990) *Late marxism: Adorno, or the persistence of the dialectic*. Verso.
Jay, M. (1973). *The dialectical imagination: A history of the Frankfurt school and the institute for social research 1923–1950*. Heinemann Educational Books.
Jay, M. (1984a). *Adorno*. Fontana.
Jay, M. (1984b). *Marxism and totality: The adventures of a concept from Lukacs to Habermas*. University of California Press.
Jewish Women's Archive. Shmuel Feiner. (Viewed on July 21, 2021). https://jwa.org/encyclopedia/author/feiner-shmuel
Moran, D. (2017). Edith Stein's encounter with Edmund Husserl and her phenomenology of the person. In E. Magri & D. Moran (Eds.), *Empathy, sociality, and personhood: Essays on edith Stein's phenomenological investigations* (pp. 31–47).
Muller-Doohm, S. (2005). *Adorno*. Polity Press.
Nenon, T., & Embree, L. (Eds.). (2013). *Issues in Husserl's ideas II*. Springer.
Paddison, M. (2004). *Adorno, modernism & mass culture*. Kahn & Averill.
Rensman, L., & Gandesha, S. (2012). *Arendt and Adorno: Political and philosophical investigations*. Stanford University Press.
Ricœur, P. (2004). *The conflict of interpretations: Essays in hermeneutics*. Continuum.
Rose G. (2014) *The melancholy science: An introduction to the thought of T.W. Adorno*. London. Verso.
Sawicki, M. (1997). *Body, text, and science: The literacy of investigative practices and the phenomenology of Edith Stein*. Springer.
Simmel, G. (1916). *Das problem der historischen zeit, Philosophische vorträge der Kant Gesellschaft 12*. Reuther & Reichard.

Stein, E. (1933) *Letter to Pope Pius XI*. English translation published by the Council of Centers on Jewish–Christian Relations; reproduced with editorial context from Carmelite Quotes and Catholic Culture Carmelite Quotes catholicculture.org.
Stein E. (1952) *On human being: A dispute between Edith Stein and Martin Heidegger* (trans. Cecily Hastings & Donald Nicholl) New York: Sheed & Ward, from Wilk, R. K. (2007). Logos: A Journal of Catholic Thought and Culture, 10(4), 104–119.
Stein, E. (1989). *On the problem of empathy*. ICS Publications.
Stein, E. (2000) *Philosophy of Psychology and the Humanities* (M. C. Baseheart and M. Sawicki, Trans.). ICS Publications.
Stein, E. (2002). *Finite and eternal being: An attempt at an ascent to the meaning of being*. (K. Reinhardt, Trans.). ICS Publications.
Stein, E. (2006). *An investigation concerning the state* (M. Sawicki, Trans.). ICS Publications.
Stein, E. (2009). *Potency and act: Studies toward a philosophy of being*. (W. Redmond, Trans.). ICS Publications.
Stein, E. (2016). *Life in a Jewish Family: An autobiography* (J. Koeppel, Trans.). ICS Publications.
Tullius, W. (2019). Time, eternity, and the transition from phenomenology to metaphysics in edith Stein, Edmund Husserl, and Eric Voegelin. *The Review of Metaphysics, 73*(2), 255–283.
Van Rahden, T. (2008). *Jews and other Germans: Civil society, religious diversity, and urban politics in Breslau, 1860–1925*. University of Wisconsin Press.
Von Boeckmann, S. L. (2004). *The life and work of Gretel Karplus/Adorno: Her contributions to Frankfurt school theory*. Doctoral Dissertation submitted to the University of Oklahoma.
Wargo, V. (2011). Reading against the grain: Edith Stein's confrontation with Heidegger as an encounter with hermeneutical phenomenology. *Journal of the British Society for Phenomenology, 42*(2), 125–138. Jackson Publishing and Distribution. https://doi.org/10.1080/00071773.2011.11006735
Waters, R., & Wright, R. (1973). *Us & them, TRO-Hampshire house publishing, from the album The Dark Side of the Moon, recorded at Abbey Road Studios by Pink Floyd*. EMI Records.
Wilcock, E. (2000). Negative identity: mixed German Jewish descent as a factor in the reception of Theodor Adorno. *New German Critique, 81*, 169–187.
Wilk, R. K. (2007). On human being: A dispute between Edith Stein and Martin Heidegger. *Logos: A Journal of Catholic Thought and Culture*, 10(4), 104–119.
Young-Bruehl, Elisabeth (2004) [1982]. *Hannah Arendt: For love of the world*, Second ed., Yale University Press.

Dr. Pól O'Hagan is an interdisciplinary researcher whose work bridges cultural policy, intellectual property governance, and immersive media. His PhD, *Credit Where It's Due: Music, Copyright and Social Authorship* (Ulster University, 2023), was awarded the 2024 Prize for Outstanding Research. He has co-authored the UKIPO's *Music 2025* report and leads innovation projects on VR archives and smart copyright management. His current research applies Edith Stein's philosophy to questions of authorship and cultural governance, advancing models of relational economics that foreground collaborative and community-based creativity.

Chapter 6
Empathy in the Teaching/Learning Relation: Stein, Husserl, and Arendt

James McGuirk

Abstract This article explores the role of empathy in the teaching/learning relationship. I argue that not only are the phenomenological analyses of empathy offered by Husserl and Stein useful for understanding that relationship, but that Stein's sensitivity to the positive primordiality of foreignness in empathy is especially illuminating with regard to the tripartite relationship of teacher-learner-world enacted in educational practice. In part 1, I introduce the problem of empathy as analyzed by Husserl and Stein and outline some differences in their respective accounts of the phenomenon in the period leading up to the publication of Stein's doctoral thesis, *On the Problem of Empathy*. While both are interested in the role of empathy in the constitution of a shared world that can become the domain of natural and social scientific inquiry, their accounts differ in that Stein's pays greater attention to the experience of foreignness as such in the perception of others. I argue that while Stein's account offers a better account of the experience of foreign consciousness, it also provides a richer account of intersubjective world constitution by resisting the principled interchangeability of conscious perspectives on the world. In the second part of the paper, I turn to the teaching/learning relationship, especially as it is described in Hannah Arendt's writings on education. I argue that Stein's account of empathy helps to make sense of several of the key features of the relation as it is described by Arendt.

Keywords Empathy · Teaching · Learning · Attention · Modelling · Formation · Husserl · Stein · Arendt

In what follows, I want to offer some reflections on the phenomenon of learning and the teacher/pupil relation and specifically how and to what extent this relation can be illumined by the phenomenological accounts of empathy found in the work of Edmund Husserl and Edith Stein. My account of the teaching/learning relationship is in large part inspired by Hannah Arendt's philosophy of education, especially as

J. McGuirk (✉)
VID University, Oslo, Norway
e-mail: james.mcguirk@vid.no

M. Lebech (ed.), *Stein and Husserl's Phenomenology of the Social in Ideas II*, Women in the History of Philosophy and Sciences 25,
https://doi.org/10.1007/978-3-032-17339-3_6

this is developed in the 1954 text "The Crisis in Education." As rich as Arendt's text is, I believe that its main insights can be deepened and supported by attention to phenomenological analyses of empathy, specifically in the work of Edith Stein. My intention is to argue not only that empathy, as developed by Stein and Husserl, is essential to the teaching/learning relation, but that Stein's specific emphasis on the irreducible encounter with otherness in empathy is crucial to making sense of the tripartite relationship between teacher, learner, and world in Arendt's account. But before addressing the question at hand, some brief comments concerning the main lines of Husserl's and Stein's respective accounts of empathy are in order.

6.1 The Problem of Empathy in Phenomenology

The problem of empathy and its role in the constitution of subjective and intersubjective world experience was a central and enduring problem for many phenomenologists, not least Husserl and Stein. It is fair to say that empathy occupies a central, if not the central, position in Stein's early philosophical writings, and while somewhat less central for Husserl, is a theme to which he returns again and again. Inasmuch as the problem of empathy is one to which Husserl often returns, it is difficult to provide a straightforward account of his view on the subject, and since the confines of the present chapter preclude a comprehensive overview of his evolving perspectives on the subject, I will limit myself to commenting on the views expressed during the period of his collaboration with Stein, especially the period leading up to the preparation of the latter's doctoral dissertation. This means treating of Husserl's work up to and including *Ideas II* and Stein's work in *On the Problem of Empathy*.[1]

6.2 Husserl on Empathy

The problem of empathy begins to emerge as a theme for Husserl around the turn of the first decade of the twentieth century. It is addressed in some detail in the so-called intersubjectivity lectures (Hua XIII–XV), briefly in *Ideas I* and then again in some detail in *Ideas II*. What is clear from these various treatments is that there is an overarching concern that enframes Husserl's engagement with the issue. This concern is the question of world constitution and the problem of founding the objectivity of science. From the time of his very earliest writings, Husserl is concerned with the

[1] Since Husserl returns to the question of empathy again and again in his later writings, it would seem unwarranted to make any claim to the effect that what appears in *Ideas II* represents his final position on the matter (see Zahavi, 2014, p. 124). This is especially true in light of Dirk Fonfara's recent re-editing of the manuscript of *Ideas II* (Husserl, 2025), which once more brings into question the exclusivity of Husserl's contribution to the original Biemel edition (Husserl, 1952). And while Stein does not return specifically to the question of empathy, her understanding of intersubjectivity remains structurally crucial to many of her later writings.

question of how something like an objective standpoint is possible, and after 1910 or so, this becomes the question of how objectivity is constituted in transcendental subjectivity.

The phenomenological reduction, which allows us to move out of the natural attitude and thematize it, is famously not a move that loses the world, but one that gains it as a residuum for constitutional analysis (Husserl, 1983, p. 113). Among the eidetic structures of consciousness that come to expression through this move are the ideality of meaning, time consciousness, the noesis-noema structure of intentionality, and also empathy. Now, while empathy is only discussed in passing in *Ideas I* (§151), it is treated in more detail in lectures from 1910–11 (*Hua XIII;* Husserl, 2006) and again in *Ideas II* (Husserl, 1989). Empathy turns out to be crucial in constituting the world as a shared world and therefore a world in which objectivity and science is possible. While all of the aforementioned structures are essential to the constitution of meaning, it is only through the mediation of empathy that an objective point of view can emerge. As such, transcendental subjectivity's constitution of the world as an objective domain of inquiry is only fully possible as transcendental intersubjectivity.

Here, it is important to be clear about what we mean by empathy. Commentators in the phenomenological tradition have often gone to great lengths to stress what empathy does and does not mean as it used by thinkers such as Husserl, Stein, Max Scheler and others (Fernandez & Zahavi, 2020; Lebech, 2015; Steinbock, 1995). In our ordinary discourse, and sometimes in scholarly debates, the term empathy is used to cover a broad range of phenomena from sympathy and fellow feeling (Bloom, 2016) to the ability to put oneself in the shoes of another (Goldman, 1970). In other words, it is used as a designator that captures the whole range of our social, psychological, and moral attunement to others (cf. Zahavi & Gallagher, 2008). But as phenomenologists have used the term, empathy is—at least initially—more modest in its meaning, designating simply the givenness of a foreign stream of consciousness in my conscious experience. What is given in empathy, for Husserl, is an alien ego. And yet, this modest or seemingly banal designation of empathy turns out to be both complex and essential for our understanding of meaning as such. For one thing, the experience of the foreign ego is not one given in direct perception, but always through the mediation of the body. It is, thus, an intercorporeal phenomenon. As a conscious subject, I find myself surrounded by other lived bodies, which I regard as bearers of I-subjects, but which are sharply contrasted with my own lived body and my own conscious first-personal experience (Husserl, 2006, p. 5). The empathic givenness is, therefore, a "quasi-perceptual presentation" (Husserl, 2006, p. 6), in which the lived body of the other is given in direct perception like other things in the world, while her first-personal subjectivity is given only indirectly in appresence (Husserl, 1989, p. 172).

But what is of greatest importance in empathy is the access it provides to another source of world experience as given through the mediation of the conscious life of the foreign ego. For Husserl (and Stein), the appresentation of the inner life of the other I-subject turns out to be enormously important for both subjectivity's constitution of itself and for the possibility of world constitution. With regard to the first point, each I experiences itself as a middle point in contrast to others which are experienced as surrounding points (Husserl, 2006, p. 6), and since the other is given in empathy as analogon to myself, his place is given as a "here" in relation to which everything else, myself included, is given as "there" (Husserl, 1989, p. 176). And secondly, inasmuch as the other is a middle point of givenness of the world, his positing of things is also potentially mine (Husserl, 1989, p. 177). This is the key role played by empathy. As Husserl says,

> I identify the thing I have over against me in the mode of appearance α with the thing posited by the other in the mode of appearance β. To this belongs the possibility of substitution by means of trading places. (Husserl, 1989, p. 177).

This substitutability is the key to establishing the objectivity of the world (Husserl, 1989, pp. 178, 240), both because it establishes my subjectivity as an objectivity in the natural world (Husserl, 1989, p. 175) and because it ensures the actualizability of my perspective on the world by the other and his by me. The mode in which the world is given (empathy rather than direct perceptual awareness) is different, but its content is the same. As Husserl says,

> Nature is constituted as intersubjectively common and as determinable objectively (exactly), and oneself as subject is constituted as a member of this objective nature. The relativity of the things of experience in relation to individual men is thus incontrovertible, and it is also beyond doubt that the positing, accomplished in empathy, of an intersubjective world in the form of natural science allows that which is intersubjectively posited to be determined "theoretically" in such a way that the content of the determination becomes independent of the individual subjects, i.e., in such a way that this content consists exclusively of determinations any subject can acquire, out of what is given to him in appearances, by means of the methodological procedure of natural science and with a sense that is identically the same for every subject involved in natural scientific research. (Husserl, 1989, p. 179).

It is through the mediation of empathy that I apprehend myself as a psychophysical unity with a perspective on the world and it is through the relation I-other that the world becomes accessible to natural scientific determination. Another way of putting the point is to say that the I grasps certain of the objects given to consciousness as other Is, and with them the plenitude of the world as common field of experience and inquiry.

6.3 Stein on Empathy

We turn now to Stein, whose differences with Husserl's views on empathy are not always so easy to discern. Independent of the question of Stein's contribution to the editing of *Ideas II*,[2] which I will not go into here, is the fact that her sustained treatment of empathy in *On the Problem of Empathy* (1916) engages only indirectly with Husserl. The reason for this is most likely that Stein understands herself as

---

[2] The matter is discussed in some detail in (Sawicki, 1997). While this is a thorny issue, it is hoped that a comparison of the new (Husserl, 2025) and old (Husserl, 1989) editions of *Ideas II* will shed light on this question.

thinking in the light of Husserlian constitutional analysis and method about empathy and so as representing an Husserlian position on the problem, which is contrasted with her critical interlocutors such as Scheler, Lipps, Geiger, and others. Husserl is not so much an interlocutor for her analysis of empathy as a starting point for her treatment of the problem.

Having said this, and in spite her failure to signal (or even unwillingness to acknowledge) them, there appear to be differences between Stein's and Husserl's early approaches to empathy (Sawicki, 1997, p. 113). It is arguable that some of this can be put down to the fact that Stein simply devotes more time to empathy than Husserl had done. The whole of the 1916 doctoral work is devoted to the problem of empathy as such and takes up social as well as epistemological aspects of the experience. Of course, empathy is discussed in some detail by Husserl in the intersubjectivity lectures and in *Ideas II*, but this is done with a very specific and narrow goal in mind; namely, the grounding of science and objectivity in transcendental subjectivity. Stein shares this interest with Husserl, but her analyses are not *entirely* framed within this concern, and so she draws out nuances in the experience of empathy that are not fully present in Husserl's account. I do not mean to suggest here that Husserl's interest in empathy is purely instrumentalist, but his treatment of the issue in these texts is more delimited than Stein's, likely because of the differences in their respective projects.

Perhaps the most significant of these nuances concerns Stein's more sustained attention to the positive role of the foreignness given in empathy. As we saw, Husserl had already described empathy as the perception of a foreign consciousness but given that he is predominantly interested in empathy as the means by which a shared world is constituted, the foreignness of the other as such tends to be passed through. By contrast, Stein, in her sustained interest in empathy as the givenness of the otherness of the other in its own right, is decidedly more attentive to the enactment of a relationship between two Is in empathy. Or, put in other (Schelerian perhaps) terms, Stein is from the beginning just as interested in empathy as an encounter between persons as an encounter across subjects with the world. That is, there is a complex, irreducible interplay between the shared and the foreign in empathy that she is trying to tease out.

Early in the 1916 work, Stein defines empathy as "the experience of foreign consciousness [as the] the non-primordial experience which announces a primordial one" (Stein, 1989, p. 14). My perception of the other in empathy is a primordial experience, the content of which is given non-primordially, as not mine. This is similar in tone to Husserl's depiction of empathy as a quasi-perceptual presentation in which the other's world experience is appresented. It also depicts empathy as similar in form to other reflexive acts such as memory and expectation, by involving primordial experience given in a non-primordial way. Thus, while the primordial experience of the I as it is given in memory is one from which the present I is temporally separated—the "now" of the remembered event was once given as my primordial "now"—the I of empathy is given as a spatially separated "now".

However, while empathy and memory are similar as to their form, they are crucially different as to their content. It is at this point that the difference with

Husserl becomes pronounced. That is to say that where Husserl saw this separation (of experience contents) as non-essential, since the point of view of the other I could be mine were the other's "here" to become mine, Stein envisions the separation of Is in empathy to be irreducible. The other's experience can never fully be mine, which is to say that the experience of the I and the other are different as to their content. It is not just that the content of the experience given in empathy is non-present in the way the experience content of a memory or an expectation is non-present. It is rather that the empathized experience can never be directly or primordially present to me. It is essentially other.

This is not to say, of course, that the content of the other's experience is simply opaque to me. Like Husserl, Stein wants to insist that the foreignness of the other's perspective becomes the ground of the communicability and shareability of the world. But it is important for her that the foreign as well as the shared is duly weighted in her account of empathy. Empathy is at once the invitation to see what the other has seen and to share the world with her, and also an encounter with the irreducibly foreign whose mystery can never be cashed out entirely in my own first-personal terms. This opacity of the other, which Husserl also acknowledges is definitive of the experience of empathy, is not pure opacity, but the positive givenness of the other as other. It is the sadness or joy of the other, for example, that first exerts a kind of gravitational pull that draws me towards the other as the experience of another centre of experience. In empathy, I am first drawn in by the content of the experience (the sadness or joy as feeling) given non-primordially, and then through this to its object towards which I am then turned (Stein, 1989, p. 12), be this the other's lost love or recent triumph. In this sense, empathy involves both the givenness of an irreducible otherness at the same time as being a co-turning towards the world. In empathy, Stein says, I achieve the standpoint of the other, but without stepping into the other's place (Stein, 1989, p. 64).

As one commentator puts it, Husserl's interchangeability of other bodies as bearers of I-subjects is replaced by Stein's sense of their irreplaceability (Sawicki, 1997, p. 112). Stein, like Husserl is interested in empathy as founding the constitution of the world as real, objective, and shared; but, by rejecting the idea of the other as an avatar for the perspective that any I could have on the world and arguing for the irreplaceability of the perspective of the other, she lays the ground for a richer account of the genesis of the common world. This turns out to be of great significance, because Stein's emphasis on foreignness as such is more than something added to Husserl's concern with the possibility of a shared world. The common world that is constituted in empathy is the world of natural scientific exploration and the social world, but it's constitution is not the result of the repeatability of points of view, but precisely their unrepeatability. It is not always clear whether Husserl, at least in the works leading up to *Ideas II*, shares this view.[3] As we saw earlier, it is often in his impersonal being that the other can co-constitute the world in intersubjectivity by actualizing an infinitely

[3] It is worth noting, in this regard, subtle differences between the older and newer versions of *Ideas II*. While the older version sometimes gestures in the direction of the plurality of perspectives as contributing positively to the constitution of the objective world (Husserl, 1983, pp. 179–180), the newer version more consistently emphasises the interchangeability of perspectives and of empathy

re-actualizable point of view (Husserl, 2006, p. 7). The foreignness of the other is a question of positionality, not of content. Or, put differently, it is the quantity of first-personal perspectives that makes the common world possible. For Stein, by contrast, the qualitative variety of world perspectives appears to make a crucial difference. By her view, there is a doubling of the content of experience as given in empathy. The irreducibility of the primordial givenness of the other and non-primordial givenness of her experience establishes a multiplicity of points of view on the world. This means that the world, in empathy, reveals itself in surprising and unexpected ways that are not only intersubjective, but interpersonal. So, empathy establishes a common world and therefore the possibility of science, but by establishing a multiplicity of irreducible centres of experience, it insists that the common world is also ineradicably plural.

6.4 Phenomenon of Learning

In the second part of this discussion, I want to explore the extent to which the foregoing analyses can help us make sense of aspects of the teaching/learning relation. Now, the first thing to note preliminarily is that the analyses of Stein and Husserl are unashamedly removed from educational or developmental settings, so that my comments here will be somewhat tentative and should be read as attempting to tease out the implications of their respective accounts of empathy for educational practice. At one point, Husserl does acknowledge that children need to be led to world-consciousness through the guidance of "mature and normal human beings who bring them up that they become acquainted with the world in the full sense of the world-for-all" (Husserl, 1970, p. 187), but this nod towards the idea of development is not given much space in *Ideas II* (Morselli, 1990, p. 321). I want to look in a little more detail at how this induction of the child into the adult world occurs and at how theories of empathy can help us make sense of this. In doing so, I want to draw on the phenomenological account of the teaching/learning relationship as it is developed by Hannah Arendt and other educational theorists influenced by her. And I want to bring these into dialogue with Stein's account of empathy in particular. In spite of the fact that Stein, like Husserl, takes her point of departure in fully formed adult subjectivity, I will suggest that there is a significant degree of coherence between her account of empathy and Arendt's phenomenology of educational practice, specifically inasmuch as both emphasize the importance of a dynamic interplay between the common and the strange in the constitution of meaning.[4]

---

as thereby rescuing the ego from the reef of solipsism (Husserl, 2025, 23). These differences are not very pronounced, but they speak to the question of Stein's contribution to the originally published manuscripts, especially given that the newer version appears to be more consistent in its emphasis on objectivity and interchangeability.

[4] In fact, Stein does return to the question of education in her 1932 work *Der Aufbau der menschlichen Person* (Stein, 2004, see also Lebech, 2015, pp. 79–99) and much of what she has to say is indeed compatible with Arendt's philosophy of education. I have chosen to omit detailed

In order to make this case, I need to first provide a brief outline of the meaning-bearing or constitutive aspects of the teaching/learning relation, as these are developed by Arendt. While there is much that could be said here, I want to limit myself to four of the central features of the relation as they are described in Arendt's 1954 essay, "The Crisis in Education".

6.4.1 Love of the World

The first of these is Arendt's description of the teaching/learning relation as fundamentally a relation of love. What she has in mind here is not first and foremost a bi-partite relationship between the teacher and the child, but a tri-partite relation in which it is precisely love of the world (*amor mundi*) that binds the teacher to the child and vice versa. What is enacted in the classroom is not merely the teacher's love of knowledge to which the child is a passive spectator, nor a contextless concern of the teacher for the child. The teacher can and should care about the child as individual, of course, but this must primarily be expressed as an interest in the world that seeks to generate the same interest in the child. The teacher, says Arendt, is qualified inasmuch as she knows the world and is able to instruct others to know it. But in so doing, she takes responsibility for the world (Arendt, 2006, p. 186) as it is and as it has been (Arendt, 2006, p. 190). Arendt describes this as an act of love. The teacher stands before a new generation as representative of her own generation and every generation before her and says "this is our world" (Arendt, 2006, p. 186). The mediation of the child/teacher relation through the world establishes an *inter-esse*, or an interest for the world that is constituted in common between teachers and pupils (Masschelein & Simons, 2013, p. 48).

6.4.2 Attention

The second feature of Arendt's thinking on the matter is that the loving core of the relation is specified through the act of paying careful attention to the world or to whatever part of the world is being studied in the particular situation. If love is the ultimate form of the shared interest in the world, then this is concretely played out in the act of paying attention. The teacher looks carefully and shows what it is possible to see and what paying attention makes possible. Educational researcher Jan Masschelein recounts the incident of a teacher charging his pupils with the task of exactly reproducing a video they have watched on YouTube (Masschelein, 2016). On the face of it, this seems a dull task indeed. Why not allow what they have watched

consideration of this later work here for the simple reason that its orientation is more general in dealing with the question of philosophical anthropology as such, while Arendt's discussion is more practically anchored in educational or classroom practice.

to inspire them to make their own video, or tell their own story? What is to be gained by simple reproduction of someone else's work? But the point is that reproducing requires careful attention to every aspect of what they have seen. They need to listen to what is said, to observe the lighting and set design, and the film's pacing. Often when we watch, we think we have "gotten the point", but there is more to be seen than what we take notice of at first glance. The purpose of this exercise was precisely to slow down, to look again, and to look better.

6.4.3 Modelling of Interest

The teacher's attention to the world, and this is the third point, is consequently a form of modelling of interest and attention. The teacher's careful attention to the subtleties of German grammar, say, demonstrate in an embodied way the kind of fine-grained attention that is possible for the child and thereby draws the latter into the same kind of attentive inquiry. As such, the business of modelling has a double orientation towards both the object of attention and the subjectivity of the child, whose attention the teacher wishes to draw in the same direction as her own. The teacher's love of the world is, then, an attending that never loses sight of itself as modelling what love of the world can look like for the child. The teacher's careful, loving attention to the world will redound to the child and generate an interest in the world in the child that manifests itself as joint or shared attention.

6.4.4 Formation

While the first three features—love, attention, modelling—speak to the form of education, the fourth—formation—speaks to its end. It also, crucially, widens the meaning of love as the ultimate characterization of the relation. If it is love of the world that binds the teacher to the child, it is also in the enactment of this relation that concern for the child as individual comes to expression. In other words, the teacher does more than simply demonstrate an inquiring stance to the learner. Rather, the very act of demonstration is itself an expression of love for the individuality and developmental possibilities of the child. Love of the world is love for the child, in other words. But this is not a love that is without tension. Arendt famously characterizes the practice of education as conservative (Arendt, 2006, p. 192), a designation that can easily be misunderstood as either a political stance or a simple resistance to change and novelty. But this is not what Arendt means. By conservative, she means instead to suggest that the teacher is primarily concerned with presenting the world *as it is* to the child. The teacher presents the world as the world that is understood by previous generations and in doing so both arrests the possibly destructive impulses of the child's spontaneity and opens up possibilities to see. She (the teacher) invites new avenues of constitution, yet does so within frameworks that constrain. Indeed, for

Arendt, it is precisely these constraining frameworks—the world as constituted—that make possible the new. The present of the classroom situation is, therefore, very much rooted in the past. But precisely in this way, it is also oriented towards the child's future. The teacher's love of and responsibility for the world, modelled through careful attention, is precisely that which will facilitate the emergence of the child as actor and thereby as one who can bring something new into the world (Biesta 2014, p. 105). In other words, it is for the sake of transformation and the new that the teaching/learning situation is conservative with regard to the world. What the teacher makes possible is the facilitation of an orientation of attention to the world in which the child becomes an addressee of the things of the world. They (things) exert a resistance to the immediacy of the will or interest of the child, and thereby facilitate the possibility of self-transcendence. The child pays attention, and in so doing becomes transformed by the being of things in the world. And through this transformation in the child, the world itself can become transformed.

We might think again of Masschelien's YouTube teacher, but the same dynamic can be played out in other pedagogical situations. Let us say that the classroom situation involves building a shelf. In the mutual attending to the materiality of the wood, the latter ceases to be an amorphous something and begins to emerge in all the specificity of its resistance and affordance. The hardness of the wood, its smoothness or roughness, and the teacher's engagement with these tell the child what is and is not possible, and these are communications to which he must attend in proceeding. Similarly, German grammar is learned in all of its intricate details, details which curtail and perhaps are initially experienced as pure resistance. But the resistance that cannot be short-circuited eventually gives way to possibilities of expression that are drawn out of the child. Or, rather, they are drawn out of the encounter between the child, the teacher, and the language while being reducible to none of these. The world opens up through German, we might say.

6.5 Empathy and Learning

In what (if any) ways can the analyses of empathy bear on these matters? At the centre of Arendt's phenomenology of the teaching/learning relationship is the idea of transformation through attending. The child has an inborn inquisitiveness and desire to know and understand, but this will not unfold of itself. It requires, to put the matter in Aristotelian or Thomistic terms, the mediation of others to actualize the potency of this curiosity (Haldane, 2009, p. 331). It is only by witnessing how the world can be attended to that the child learns how she might attend to the world. Husserl acknowledges this when he speaks about the child requiring guidance on the way to mature acquaintance with the world. But something more is required, to which Husserl's account of empathy provides little answer. This something more needs to be part of the account of empathy. In order for the child to be led, she must first be drawn into a relation with an other who is apprehended as a possible model for world-orientation. The child needs to be drawn to the other in something like admiration in order to get things moving in the first place.[5]

Crucially, the empathic perception that Arendt has in mind in her account is one in which the child perceives the teacher not as "one like me" or "one who sees what I would see in her position", but precisely one who sees something I have not seen. The child's perception of the teacher's way of attending is at once the perception of a mysterious way of seeing the world and an invitation to let the world address her (the child) in this unexpected way. What the child perceives exerts an attractive force that draws her to the teacher, and through the teacher to the world as domain of objective inquiry. This brings Arendt into a more fruitful dialogue with Stein than with Husserl. For all three, the empathic relationship between the two unfolds in terms of a shared focus on the world (Stein, 1989, p. 10), but for both Arendt and Stein, this is more complex than a mere confirmation of points of view. Rather, the child needs to be drawn into the common world by a perception of the experience of the other as alien and enigmatic. And in this case, what is given is also a sense of admiration for the kind of world orientation the teacher models. The teacher's perspective exerts a double gravitational pull on the child, not only to what she (the teacher) has seen, but to her capacity to attend to the world in the first place.

Stein, we remember, distinguishes between the direct or primordial perception of the experiencing life of the other and the irreducibly non-primordial perception of its content. And while admiration is by no means a necessary feature of empathic perception for Stein, she does insist that empathy is led by the other rather than projected (Stein, 1989, p. 10). I am drawn out of myself by the experience of the foreignness of the other ego. We see a similar structure in the teaching/learning example in the form of the double aspect in the focus of the child, who attends to the world in ever more careful detail and to the teacher's attention to the world. The child is drawn both to the world as a domain of inquiry and to the teacher as a centre of experience. She is drawn to attention to the world by being drawn in admiration to the model that shows how this is possible. Prior to being transformed by the resistance of the world, the child is transformed by her witness of a possibility to be towards the world, which he initially encounters as both puzzling and appealing. While one of these foci is primordial and the other is not, it would be a mistake to think of them as chronologically separated. It is not the case that the child is first drawn to the teacher and only thereafter to the world. The relation must, as Arendt insists, remain integrated and tripartite.

In Stein's terms, this amounts to the personalist core of education to which she returns in the later work (Stein, 2004). It is in the interpersonal encounter that the possibility of education manifests itself. Shared world constitution is a central outcome of this encounter, but this is important because it is an unfolding of the child's possibility to be. In this sense, the epistemological achievement of empathy

[5] The notion of pedagogy as drawing the child out of himself through the experience of admiration of another is as old as Aristotle's account of moral education in the *Nicomachean Ethics*. It has recently been restated in the work of Linda Zagzebski (Zagzebski, 2019).

takes on a thoroughly personalist tone in a way that was hinted at but never developed by Husserl.

But might this not be just a developmental issue? That is, while a theory of empathy might usefully be applied to the specificities of the teaching/learning relation, the developmental context of the latter makes it unlikely that any line of influence could proceed in the opposite direction and that the teaching/learning relation could tell us anything about empathy as such. One can imagine the objection, for example, that while the child's attention must initially be scaffolded by the teacher in order to reach maturity, her focus on the teacher's experiential flow should be understood as a place-holder that will fall away in time, so that mature world constitution will follow the Husserlian schema. But this does not seem to follow from the analyses of Stein and Arendt. Learning to attend properly to the world requires the mediation of role models, especially for the child, but their arguments also suggest that this process trains the maturing and the mature subject to attend to the world with others. That is, we are not simply brought into the world by others who then withdraw, but are taught how to attend to the world as a shared field of experience. Learning can be as much a feature of adult existence as of the world of the child, and learning will always involve the capacity to attend to the attention of the other whose capacity to see strikes us as salient and worthy of imitation. But beyond even the business of learning, our everyday world experience is deepened in and through our awareness of the experience of others. This is, in fact, why we grieve the loss of family and friends. It is right to say that we grieve the loss of the person, but at a deeper level, this means that we are grieving the loss of a point of view on the world, which made our own world experience more nuanced and profound (Sokolowski, 2000, p. 154). Others are not first and foremost guarantors of the credibility of our experience, but companions in the disclosure of the world. Our experience becomes richer when it is shared and when the process of sharing is manifest as both a holding in common of the world or object under consideration and attention to the uniqueness of the co-attending other or others.

6.6 Conclusion

The teaching/learning relationship is one in which we learn to attend to others in attending to the world. The teacher's position neither effaces the uniqueness of the child's I, nor is effaced by it. Rather, the uniqueness of the child's I is cultivated by the relation of modelling. It is precisely this that draws the child out as subject (Biesta, 2014, p. 126) and co-subject. To share the world in empathy entails both a joint attention to the world or whatever object is under investigation, and an attention to the unique otherness of the other's attention. Following Stein's analysis of Lipps' acrobat (Stein, 1989, pp. 16–17), we might say that empathy makes possible a shared world without the complete overlap of I positions, or better, that the world is truly shared only to the extent that there is no complete overlap. The child does not wholly re-enact in inward experience the point of view of the teacher anymore than the spectator does with the acrobat; in part because the I never forgets itself fully in its engagement with the object of its attention, and in part because the sharing of the world remains irreducibly empathic, which means that the perception of the otherness of the other remains a constitutive part of this experience. As such, the child not only learns to attend to the world in the teacher/learner relation, but learns to appreciate and agree to "a world with multiple centres" (Sawicki, 126).

All three of the thinkers we have been considering—Husserl, Stein, and Arendt—are interested in the constitution of the world as a shared domain of inquiry and action. For both Husserl and Stein, empathy is crucial to this process, but they differ in their apprehension of the extent to which the substantive foreignness of the other I is itself constitutive. In mediating this discussion, Arendt has proven useful through her analyses of the teaching/learning relation, which seems to suggest that at the developmental stage and beyond, our capacity to attend to the world as a shared domain of inquiry is ineluctably tied to our capacity to attend to the unique way in which foreign centres of experience open this world.

References

Arendt, H. (2006). The crisis in education. In *Between past & future* (pp. 170–193). Penguin Books.

Biesta, G. (2014). *The beautiful risk of education*. Routledge.

Bloom, A. (2016). *Against empathy*. Random House.

Fernandez, A., & Zahavi, D. (2020). Basic empathy: Developing the concept of empathy from the ground up. *International Journal of Nursing Studies., 110*. https://doi.org/10.1016/j.ijnurstu.2020.103695

Goldman, A. (1970). *A theory of human action*. Prentice Hall.

Haldane, J. (2009). *Understanding education. Practical philosophy: Ethics, society, and culture*. Imprint Academic.

Husserl, E. (1952). *Ideen zu einer reinen Phänomenologie und phänomenologischen Philosophie, Zweites Buch* (M. Biemel,.Ed.). Martinus Nijhoff.

Husserl, E. (1970). *The crisis of European sciences and transcendental phenomenology*. Northwestern University Press.

Husserl, E. (1983). *Ideas pertaining to a pure phenomenology and to a phenomenological philosophy. First book: General introduction to pure philosophy* (F. Kersten, Trans.). Springer.

Husserl, E. (1989). *Ideas pertaining to a pure phenomenology and to a phenomenological philosophy: Second book: Studies in the phenomenology of constitution* (A. Schuwer & R. Rojcewicz, Trans.). Springer.

Husserl, E. (2006). *Basic problems of phenomenology* (J. Hart & I. Farin, Trans.). Springer.

Husserl, E. (2025). *Ideen zu einer reinen Phänomenologie und phänomenologischen Philosophie. Zweites Buch: Phänomenologische Untersuchungen zur Konstitution und Wissenschaftstheorie sowie das Nachwort zu meinen Ideen* (D. Fonfara, Ed.). Springer.

Lebech, M. (2015). *The philosophy of Edith Stein*. Peter Lang.

Masschelein, J. (2016). The Language of the School: Emancipating or Alienating? Keynote Address at Workshop on *Education, Civic Virtue, and Citizenship*. Nord University. 7 December 2016.

Masschelein, J., & Simons, M. (2013). *In defence of the school: A public issue*. E-ducation, Culture, & Society Publishers.

Morselli, G. (1990). Husserl, child education and creativity. In A.-T. Tyminiecka (Ed.), *The moral sense and its foundational significance: Self, person, historicity, community, AHua XXXVI* (pp. 321–327). Kluwer.

Sawicki, M. (1997). *Body, text, and science: The literacy of investigative practices and the phenomenology of Edith Stein*. Springer.
Sokolowski, R. (2000). *Introduction to phenomenology*. Cambridge University Press.
Stein, E. (1989). *On the problem of empathy, CWES III, ICS*. Publications.
Stein, E. (2004). *Der Aufbau der menschlichen Person, ESGA XIV*. Herder.
Steinbock, A. (1995). *Home and beyond: Generative phenomenology after Husserl*. Northwestern University Press.
Zagzebski, L. (2019). *Exemplarist moral theory*. Cambridge University Press.
Zahavi & Gallagher, S. (2008). *The phenomenological mind*. Routledge.
Zahavi, D. (2014). *Self and other*. Oxford University Press.

James McGuirk is Professor of Philosophy and Director of the *Centre for Diaconia & Professional Practice* at VID Specialized University, Norway. He is formerly Director of the *Centre for Practical Knowledge* at Nord University, Norway. He is co-editor of *Philosophy of Improvisation: Interdisciplinary Perspectives of Theory and Practice* (2021, Routledge) and has published widely on themes in Phenomenology – especially habit, improvisation, and value, Practical Knowledge, and the interface between Philosophy and Professions research. Current projects include values and critical thinking in schools, and the role of literature in Professions education.

Part III
Discerning the Distinct Positions of Husserl and Stein

Chapter 7
Stein's Concepts of "Lifepower" and "Core" in the Description of the Living Body

Valentina Gaudiano

Abstract The question of the body has been central to phenomenological research and, starting with Husserl's early reflections, has moved almost all of his pupils of the first and subsequent generations to ever more in-depth research on the subject. Edith Stein was also strongly attracted to this question, making it the starting point for a genuine anthropology. In this contribution, I would like to look at the question by referring to the historical period in which Stein and Husserl collaborated strongly and, above all, in which Stein was able to draw heavily on the continuous evolution of her master's thought. The many connections between the two thinkers will emerge, as well as the originality of the young student who, following her doctoral work, embarks on an autonomous path that—through some concepts such as "lifepower" and "core", at which we shall look closely—shows her originality regarding the analysis of the body and the person.

Keywords Living body · Empathy · Causality · Person · Intersubjectivity · Life power · Core

7.1 Introduction

It is well known and accepted among scholars of Stein that the center of her research, beginning with her doctorate dedicated to the issue of empathy, is the human person, and that the results achieved in her thesis are nothing more than the beginning of an in-depth analysis that moves increasingly from an anthropological perspective to better define the characteristics of the human person while maintaining a strong phenomenological perspective, even when it meets Christian metaphysics and opens up to its contributions.

V. Gaudiano (✉)
Sophia University Institute, Florence, Italy
e-mail: valentina.gaudiano@sophiauniversity.org

M. Lebech (ed.), *Stein and Husserl's Phenomenology of the Social in Ideas II*, Women in the History of Philosophy and Sciences 25,
https://doi.org/10.1007/978-3-032-17339-3_7

The primary source of her research, albeit complemented by influences and inspirations by several of her colleagues as well, is and remains the Master, Edmund Husserl, with whom she maintains a not always straightforward relationship.[1] In fact, the work of Stein that emerges is marked by this relationship of admiration and distancing while in search of its own identity and originality. And it is evident that the period that we consider in the context of our study—1916–1925—is precisely the period that most deeply marks Stein and that gives the philosophical imprint that will be maintained throughout her intellectual journey. These are, in fact, the years in which Edith Stein concludes her work on empathy—particularly appreciated by the Master—and devotes herself to the development of her anthropology, and almost at the same time to the tiring and intense work of being Husserl's assistant, with the task of organizing his manuscripts of *Ideas II*.[2]

Stein dealt with the question of the living body in *On the Problem of Empathy*, and deepened it in the following years in *Philosophy of Psychology and the Humanities* as well as in the lessons of *Einführung in die Philosophie*. It seems to us that it is here that the pupil shares the most with the teacher. Therefore, we want to look at these Steinian writings in relation to *Ideas II* to highlight, precisely, the common traits and divergences that already appear, as well as the appearance of the concept of "lifepower" and "core" not present in Husserl—at least not in said book or in the years being considered[3]—and which clearly refer to the priority of the living body over the transcendental one.

7.2 The Constitution of the Living Body

Deeply impressed by the lectures on *Natur und Geist* given by Husserl in 1913–15 in Göttingen, Edith Stein felt moved to deepen these themes through a study of empathy. While Husserl devotes much time to this topic in *Ideas* and in the *Cartesianische Meditationen*, he lacks an accurate description of the empathic phenomenon, its characteristics and differentiation from similar phenomena, which we find instead in Stein's writing.[4] As Sawicki says, "Husserl employed the term 'empathy' *(Einfühlung)* in provocative, diverse, and inconsistent senses" (1997, p. 89). His attention

[1] In both the biography and in various letters, we can read about the fluctuating relationship between Stein and Husserl and about the different viewpoints of both. Stein looked at herself as an assistant, who worked autonomously; Husserl wanted someone to help in his own work. See: Stein (2001), 2003).

[2] It is not our primary interest here to go into Stein's work on the Husserlian manuscripts, specifically on the edition of *Ideas II*. Stein speaks about it in many letters, especially to Roman Ingarden and in her biography (Ingarden, 1962; Stein, 2000, 2001), and there are several studies on this subject, including the last edition of the *Ideas II* in German, such as from Imhof, Varga, Calcagno, just to name a few. See: Imhof (1987), Sawicki (1997), Calcagno (2006), and Varga (2016).

[3] We can find, for example, the use of the term lifepower in *Die Lebenswelt* (Husserl, 2008).

[4] Stein herself points this out in relation to her choice of topic for her doctoral thesis: "Husserl in his course on Nature and Spirit had maintained that an objective external world can only be experienced intersubjectively [...]; which means that such an experience presupposes other individuals. Husserl,

seems to fall, as Manganaro (2000, p. 103) points out, much more on the question of intersubjectivity as a character proper to the human community as well as an element of the constitution of the world as a world of values, a world of culture. Stein "describes its essence, she identifies its genesis and structure, she scans its mode of implementation" (Manganaro, 2000, p. 103). I would almost say that her study does not want to be merely "instrumental or preparatory" to intersubjectivity, as one could think by reading the Husserlian analysis.

In *Ideas II*, the space dedicated to empathy as such is relatively limited, compared to other themes addressed there (the constitution of the Ego and the distinction between the living body and the object body), despite the fact that the lectures that had led Stein to deepen the concept of empathy—*Natur und Geist*—are those that initially converge in *Ideas II*. In effect, however, the recovery of the intersubjective dimension and the development of the discourse of a common spirit as the very foundation of the world to which one belongs both seem to become central for Husserl.

This seems to me to speak in fact of a first point of divergence between Husserl and Stein, not yet at a proper phenomenological level, but one of intent. The first Steinian study, as well as her interventions in the drafting of *Ideas II* as in her contemporary works, show the clear Steinian intent to focus on the human person, while Husserl worked, first of all, on the foundation of philosophy as a science.[5] Letters and notes of that time even show the intent of a broad project of elaborating an ontology of the

following Theodor Lipps, named this experience 'empathy' but did not explain what it consisted of. Here there was a gap worthwhile filling; I wanted to explain what empathy meant" (Stein, 1986, p. 269).

[5] To confirm what has been said here, we recall the choice of empathy as the first aspect of phenomenology to be explored by Stein. Empathy, in fact, is understood as a way of accessing not so much and not only otherness, but one's own subjectivity: it is through empathy, or rather by recognizing a living body in the other, that we learn to recognize ourselves as such. Secondly, we can look at the progression of her reflections with a focus on the psyche and community dynamics, continuing through to the analysis of the state; all themes inherent to the human person. Husserl's phenomenological analyses, on the other hand, are strongly centered on the theme of the ego in order to redefine the terms of the cognitive relationship and access to the world, not so much the specificity of subjectivity—which is also considered and explored by the Master, but only secondarily and in later years (think, for example, of the manuscripts on intersubjectivity that were published posthumously, as well as the Fifth Meditation). We read in a letter to Roman Ingarden how Stein and Husserl, precisely during the revision of *Ideas II*, moved on divergent fronts: "He is now concerned with the constitution of nature (naturally without taking into account the elaboration), while I am continuing with the elaboration, which I have authorized myself to do without encountering any opposition, and have now reached the person. The natural consequence is that we hardly talk to each other anymore" (Stein, 2001, 28.01.17). In another letter, also to Ingarden, Stein explains her interest in the person, evoking the need to delve deeper into religious and metaphysical themes: "It is impossible to complete a doctrine of the person without addressing questions about God, and it is impossible to understand what history is. Of course, I don't see it clearly yet. But as soon as the *Ideas* are ready, I would like to approach these things. These are the questions that interest me" (Stein, 2001, 20.02.17). Finally, she confirms her intention to study the person in a letter dated 1918, in which she refers to her work on psychic causality.

person not then carried out in that form, but only in and through the published works.[6] In a certain sense, "Stein exhibited, young as she was, a marked independence in appropriating Husserl's ideas in her own way and for her own purposes" (Baseheart, 1997, p. 22).

But let's go back to the Steinian study of empathy, which requires the transition from the pure self to the constitution of the living body.

The structure of the chapters dedicated to the constitution of the psychophysical individual explicitly anticipates that of *Ideas II*, in which Husserl, however, starts from the constitution of the animal nature—an aspect that we find in the next work of Stein (*Philosophy of Psychology and the Humanities*).[7] From the pure self, therefore, one passes on to the soul and to the constitution of the living body, and then to the analysis of empathy. In conclusion, Stein dedicates a chapter to the spiritual sphere, understood through empathic acts. Even in *Ideas II* there is space for the spiritual, but certainly the Husserlian analysis is much wider and more articulated than the Steinian one—which, having empathy as its central object, can only limit the other parts of the research.

Not only the structure of the analysis, but the very descriptions of the living body present common main points and terminology. The following is an exemplification of these common aspects:

> thus we must say, more precisely, that every other object is given to me in an infinitely variable multiplicity of appearances and of changing positions, and there are also times when it is not given to me. But this one object (my physical body) is given to me in successive appearances which can vary only within very narrow limits. As long as I have my eyes open at all, it is continually there with a steadfast obtrusiveness, always having the same tangible nearness as no other object has. It is always "here" while other objects are always "there" (Stein, 1989, pp. 41–42).

That seems an echo of what Husserl says: "while I have the freedom to change my position towards all other things at will, […] I do not have the possibility of removing myself from my body or it from me"[8] (Husserl, 1952, p. 159).

This belonging cannot be constituted only in an external perception, as a proper body perceived externally is also an absolutely unique body, but not "my own body". If we think about sensations, we realize that, even if they are given in an absolute way, they do not spring from the pure Ego; they do not assume the shape of the cogito, because they are always found in a spatially localized "where". These

---

[6] Stein mentions this both in her doctoral work and in a number of letters, including one of 09.03.18 written to Fritz Kaufmann: "So I am now working on the analysis of the person I told you about" (Stein, 2000). As the editor points out in commenting on this statement, the work on the person probably flowed into the immediate works, such as *Philosophy of Psychology and the Humanities*, and then into the much later published works on anthropology (ESGA IV, and V).

[7] Indeed, it is precisely some of the parts of *Ideas II* that seem to date directly back to Stein's intervention in the reconstruction and arrangement of the master's manuscripts (Hackermeier, 2008, p. 84ff).

[8] „Während ich allen anderen Dingen gegenüber der Freiheit habe, meine Stellung zu ihnen beliebig zu wechseln, […] habe ich nicht die Möglichkeit, mich von meinem Leibe oder ihn von mir zu entfernen".

"wheres" are gathered together in the unity of my body precisely by being placed at different distances from me. I relate all the parts of my body to me in a "zero point of orientation" that my own body surrounds.

This body, however, is not always in a situation of stillness; if it moves in space, it presents the experience of "I move" with the characteristic consequence that if it moves it changes the image of the outside world. Expressive phenomena are manifestations of the sensate life of the individual. Even experiences of the will are fundamental in the constitution of the psychophysical individual because this one depend on them, on the subjects and their bodies.

All this, however, does not yet say everything about the human being. It is only in the empathic act, in putting oneself in the place of the other, that I come to find myself in that "there" that is my possible "here", obtaining a new orientation in a non-original way. By empathizing with an external body, we recognize in it a bearer of sensate life, since the expressions of the face or a certain way of behaving manifest the outer side of an inner experience. In this way, empathy firstly serves as constitution of self, and secondly as constitution of an intersubjective world. We find this in Stein as in Husserl, at least in certain passages of *Ideas II*, such as the following: "Only with empathy and with the constant direction of the observation of experience towards the life of the soul appraised through an extraneous body and taken objectively together with the body does the completed unity of a human being constitute itself, and do I further transfer this to myself" (Husserl, 1952, p. 167).[9]

The closeness of Stein to the Master is very evident in these passages; this is the basis on which she identified intersubjectivity as a condition of possible knowledge of the existing external world.[10] The remark on the essentiality of the body as a constitutively necessary element for the individual is clear, but stands in contrast to the Husserlian position in this regard.[11] Where Husserl, for example, emphasizes the interchangeability of bodies, Stein emphasizes their irreplaceability. And the "need of the viewpoint of the other in order to complete my apprehension of my own body"

---

[9] In Stein, we find a more concise passage that expresses the same conclusion: "Now, as we saw on a lower level in considering the living body as the center of orientation, the constitution of the individual different from self was a condition for the full constitution of our own individuality. Something similar is also found on higher levels" (Stein, 1989, p. 88).

[10] Already in the *Ideas II*, Husserl expresses himself in these terms regarding the necessity of empathic experience for the constitution of the other as I-man: "Only with entropathy (another translation of the term empathy), with the constant directing of experimental observations towards psychic life represented together with the extraneous to proper body, and always taken objectively together with the proper body, is the concluded unity of a constituted human being, a unity that I then transpose onto myself". (Husserl, 1952, p. 560). But we also find something in this direction in the *Cartesianische Meditationen* (Husserl, 1950, p. 136).

[11] Below is what Stein wrote to Ingarden on the subject of constitution, linked to that of empathy: "Incidentally, following this, I suddenly had a breakthrough, after which I imagine I pretty much know what constitution is—but in a break with idealism! An absolutely existing physical nature on the one hand, and a subjectivity of a certain structure on the other, seem to me to be prerequisites for a vivid nature to be constituted. I have not yet gotten around to confessing this heresy to the master" (Stein, 2001, 03.02.17).

is not enough for Stein, who shows “that particular individual bodies are necessary for me to know other particular human beings” (Sawicki, 1997, p. 112).

7.3 The Development of the Sensate Dimension

So far, Stein has focused on the cognitive bridge of otherness that is revealed to be an essential channel for a fuller and deeper knowledge of oneself. In *Philosophy of Psychology and the Humanities*, published in 1922, the question of corporeity is expanded to a comparison with psychology to better clarify the sensate and mental aspects that had emerged from Husserl’s phenomenological analyses—, which represent, according to Stein, an essential source for her work.[12]

Stein, unlike her teacher, therefore follows a path that leads her from the analysis of the lived experience relative to the knowledge of the other to the constitution of the human being as a sentient and mental being, arriving at the intersubjective dimension that manifests itself in human associative forms. Although at first glance one might get the impression that Stein follows Husserl in thinking that the ego is the first evidence given to us, a pure “I” as a stream of consciousness, subject of experience, without quality, in reality, for her, it does not exist as such, except in its intimate and essential link with the body.[13] We can say, with Beckman-Zöller, that “she takes consciousness as a starting point in the full Husserlian sense, and with this proposes a clear example of phenomenological analysis of consciousness or of its correlates in the manner of her colleagues of the circle of Göttingen-Munich” (2010, p. XXXIII). But she does so with her distinctive style, searching for those insights that from her perspective were lacking in the analysis of the psychic sphere, affirming:

> We are deviating a little here from the presentation of [Husserl’s] *Ideas* in our version of the concept of consciousness. His presentation was oriented essentially toward the world of the natural attitude, and it included as consciousness all manifolds that constitute this unity: the noetic and the noematic alike. If we now separate consciousness in the sense of the noetic from the correlates of all the steps, this seems to us to be required by Husserl’s own investigations concerning original time-consciousness, and we hope to find his concurrence in it (Stein, 2000, p. 8).

Here lies a second element of distance between Husserl and Stein: while the first, within *Ideas II*, persists in keeping existence separate from the transcendental sphere of the analysis of consciousness (even when such analyses obviously refer to the

[12] “I’ve been helping Professor Husserl for nearly two years with the preparation of large publications. During this time, all his manuscripts from the last years have been at my disposal (among them also those that have to do with the topic of psychology and the humanities) and it goes without saying that important influences on my own work came out of the stimulation that I was receiving in this way and in many conversations” (Stein, 2000, p. 2).

[13] Empathy cannot be explained for Stein without the body constituted as a living body. After this analysis, it will be definitely clear that constitution of oneself as a human being is not possible without a living body.

contents of consciousness within which reality appears), Stein believes it is necessary to link the analysis of consciousness to the reality of the psychophysical subject by analyzing it. In fact, in *On the Problem of Empathy*, she had already run into the impossibility of proceeding with her analysis while remaining in the sphere of the pure self, considered a fiction. The reality of our experience is linked to the body and requires us to always consider the self as a psychophysical self. In *Philosophy of Psychology and the Humanities,* starting from the observation of the lived experiences and their continual appearance in new or old lived experiences, Stein realizes that there is a certain influence on the general state of the stream of consciousness.[14]

Therefore, we can speak of a form of causality that exists between the flow of lived experiences and the "incursions" of the vital sphere of experience: something happens by intervening on the vital feeling and in general on the lived experience. As she writes in *Philosophy of Psychology and the Humanities*:

> Each shift in the sphere of "self-discovery," as we were saying before, or of life feelings (as we wish to say now with regard to the role that they play) requires a shift in the total course of simultaneous experience. If I feel myself to be weary, then the current of life seems to stagnate, as it were. It creeps along sluggishly, and everything that's occurring in the different sensory fields is involved in it. The colors are sort of colorless, the tones are hollow, and every "impression" – each datum that is registered with the life stream against its will, so to speak – is painful, unpleasant. Every color, every tone, every touch "hurts." If the weariness subsides, then a shift enters the other spheres as well. And in the moment where the weariness changes into vigor, the current starts to pump briskly, it surges forward unrestrainedly. Everything that's emerging in it carries the whiff of vigor and joyfulness" (Stein, 2000, p. 13).

Through the analysis of sentient causality, Stein develops some concepts which we cannot find in the *Ideas II,* the first of which are vital feelings and vital states:

> The two expressions don't signify the same thing. A conscious being's states of living are wont to assert themselves in the manner peculiar to consciousness. Such a consciousness of life's status, its being experienced, is a feeling of living. But it's also possible that living states emerge without manifesting themselves in feelings of life (Stein, 2000, p. 21).

According to Stein, the succession of changing vital states indicates a greater or lesser *lifepower* which is a persistent real quality of the ego. But with this we are in the realm of the real ego, the sentient one, that possesses certain real qualities and

[14] "Determinations of a reality, its statuses and properties, manifest themselves as immanent contents in the life feelings – just as in the extra-egoic data […]. A momentary determination of my ego – its life-status – manifests itself in the life feeling, and in turn such determinations manifest an enduring real property: lifepower. Of course, the ego that is in possession of this real property shouldn't be confused with the pure ego originally experienced as point of radiation of pure experiences. The ego is grasped only as a bearer of its properties, as a transcendent reality that comes to givenness by manifestation in immanent data but never becomes immanent itself. We shall designate this real ego, its properties and statuses, as the sentient. We now see that consciousness and sentience are distinguished from one another in their basic essences: consciousness as realm of 'conscious' pure experiencing, and sentience as a sector of transcendent reality manifesting itself in experiences and experiential contents. Transcendent reality comes into consideration for us at this point only insofar as our investigation of causality is concerned, and accordingly, insofar as we're compelled to continue our investigation upon this new ground" (Stein, 2000, p. 20).

that cannot be confused with the pure ego, which is originally experienced as the radiation point of the pure lived experiences.

At this point, we stress that on the common field of the sentient experience Husserl speaks about states—*Zustände*—and characteristics—*Beschaffenheiten*—but not as "life states". This is a word that makes a difference in the question of the sentient between Stein and Husserl: the appearance of the life-aspect, which is only used in Husserl when he reflects on the *lifeworld/Lebenswelt*. On this point, the divergence with the teacher emerges that Stein in honesty had anticipated at the beginning of the discussion: that is, her decision to move on to the level of reality as soon as she experiences it, thus abandoning the eidetic plane on which, instead, Husserl insists, as we read in the following passage of *Ideas II*:

> The study of such mental states, called perceptions as states of real individuals in the real world, is a matter for psychology, the science of experience. However, perceptions cannot only be investigated as actually existing states in the context of factual units of consciousness, belonging to factual psychophysical individuals in the factual world, regardless of whether in singular individual cases or in experiential science generality; rather, we can carry out an "eidetic reduction", eliminate all question of real existence, all judgement of such and fulfil the attitude of purely eidetic research (Husserl, 2025, p. 104).[15]

By contrast, the Steinian discourse insists on carrying forward, in parallel, the phenomenal and the real plane:

> Now, where we have found a real substrate of effecting, naturally it can't include any pure experience or, accordingly, any moment of experience as a segment in the causal occurring. Reality exercises no effects upon pure experiencing. However, the experiences themselves as well as their contents, to some extent, are manifestations of real conditionalities and properties such as the life feelings (Stein, 2000, p. 49).

The persistent qualities of the real ego appear as substrates of sentient causal happening: in practice, providing or taking away energies from the lifepower[16] is the cause of the sentient process as a change of other sentient qualities. Therefore, everything that comes into existence in the sensate states derives its energies from the lifepower. Sentient living appears as a realization of the lifepower and without this it would not be thinkable; it follows that one cannot think of the sentient life without causality. Of course, one cannot here understand sentient causality in the

---

[15] „Die Erforschung von solchen psychischen Zuständen, genannt Wahrnehmungen als Zuständen wirklicher realer Individuen der wirklichen Welt, ist Sache der Psychologie, der Erfahrungswissenschaft. Wahrnehmungen können aber nicht nur als tatsächlich existierende Zustände im Zusammenhang faktischer Bewusstseinseinheiten, zugehörig zu faktischen psychophysischen Individuen in der faktischen Welt erforscht werden, gleichgültig ob im singulären Einzelfall oder in erfahrungswissenschaftlicher Allgemeinheit; vielmehr können wir eine ‚eidetische Reduktion'vornehmen, alle Frage nach realem Dasein, alle Urteilssetzung von solchem ausscheiden und die Einstellung rein eidetischer Forschung vollziehen".

[16] This concept represents one of the relevant aspects of Steinian anthropology to which little attention has been paid. Christof Betschart has given an initial insight into this concept (2009, pp. 154–184; 2010, pp. 33–64; 2017, pp. 224–225). Emanuele Caminada also refers to the use of this concept in Stein to clarify her searching for an explanation of the different forces which are present in the soul (2019, Footnote 58, p. 203).

same way as mechanical or physical causality. In the latter, in fact, the unity of causal happening is exercised through the unity of material nature from which individual things emerge as centers of happening. In sentient causality, by contrast, we are limited to the sentient states of an individual who would correspond to matter and in which qualities emerge as individual analogous centers of things.

The general causal law, according to which "every sentient event is determined in a causal way", is not sufficient to clarify the question of the determination of the sentient. The ability to receive sensitive data, but certainly not as certain data, may depend on my level of lifepower. The causal laws of the sentient do not prove certain but probable, even if they can express obvious connections.

When experiences emerge from the flow characterized by moving towards something to which we refer (intentional experiences) we are faced with apprehensions and acts. With these, the mental life begins; the ego's gaze can be directed to any object or a series of objects, and thus have a succession of apprehensions of single images and a continuous apprehension that takes place in apperception, synthesis and motivations.

Now, in order to give an act to the ego, a certain amount of lifepower is necessary. There are acts—such as thinking or remembering, among others—that are carried out according to a momentary feeling; others participate in a particular way in the vital feelings.

In the vital sphere, it is necessary to distinguish between a sensitive degree and a mental one as different roots of the sentient, since when it comes to sensitive lifepower the psyche seems rooted in the material nature, while the mental lifepower depends on it. In addition, the mental power presents itself as receptivity to the objective world, which causes impulses to flow in.

So, "the life of sentience appears to us, in our investigations, as a result of the co-operation of powers of various kinds" (Stein, 2000, p. 115). With the *sensory* lifepower, it takes root in the world of nature from which it takes strength. With its *mental* lifepower, the world of objects becomes accessible to the psyche.

Returning now to the comparison between Stein and Husserl, we can say that the course of the discussion on sentient causality and motivation, as well as their close interconnection, does not show particular divergences with the position developed by Husserl in *Geist und Natur* and the subsequent elaboration of *Ideas II*. Nevertheless, we find some divergences: the development of the vital aspect, present in Husserl only later, in the context of the Lifeworld; the connection between consciousness and the psychophysical individual; and Stein's primarily anthropological interest. However, we also encounter a completely new concept, foreign to Husserl: that of *Lebenskraft*/lifepower. It represents the newest element that Stein brings to the analysis of the living body and the person, even in these early expressions of her philosophical thought.

7.4 The Novelty of "Lifepower"

In explaining the central aspects of human sentient reality, we have repeatedly used this term which, in fact, plays a central role not only in the first part of *Philosophy of Psychology and the Humanities*, but also in the second—the one dedicated to the community.

It is a concept that, linguistically speaking, can lead to misunderstanding—being today particularly linked to areas such as spiritualism, vitalism, New Age etc., areas not strictly belonging to the complex of philosophical reflections. Stein refers only to the inspiring source of this concept and the subsequent analysis of the sentient in the Lippsian *psychic power*, whose content is briefly mentioned in a footnote (34) of *Philosophy of Psychology and the Humanities*. Her choice not to adopt the same terminology of Theodor Lipps probably indicates a need for Stein to also distance herself from his analyses in the development of an autonomous reading of sensate life.

Stein introduces this term in the first pages of the chapter on "sensate reality and causality", after making the mentioned distinction between *feelings of living* and *states of living*. Lifepower is an "enduring real property" (Stein, 2000, p. 22) which manifests itself in the vital feeling, in the changing of characteristics of my ego. It concerns, therefore, the different vital states; that is why the lifepower concerns exclusively the real ego and not the pure one.

As Betschart has already pointed out in his own work on lifepower (2009, 2010), it is important to clarify that when discussing this aspect of the psyche, Stein is not referring to mysterious forces or something esoteric, but rather seeks a way to give voice to that specific aspect of the human being which is the activation of vital feelings. Lifepower, in fact, manifests itself within the life sphere of the human being and contributes, together with the concepts of sentient causality and motivation—which are present in the Husserlian analyses—towards explaining the mechanisms of the psyche and thereby of the human being, since in all three aspects both the body as psychophysically intended and the mental soul are involved. The same concept of lifepower is distinguished by Stein into two types, sensitive and mental, and thus becomes the concept with which she summarizes all the results of research conducted in *Philosophy of Psychology and the Humanities*. We read in the final paragraph:

> the life of sentience appears to us, in our investigations, as a result of the co-operation of powers of various kinds. We distinguish a sensory lifepower, which converts itself into the reception of sensory data (into different capacities for the reception of sensory data, that is) as well as into sensory impulses and their activities. Besides that, *sensory* lifepower serves for the conservation of *mental* lifepower, from which the mental activities and capacities are fed (Stein, 2000, p. 115).

With the introduction of lifepower, we are at the heart of the Steinian analysis of the person. Through the double concept of lifepower, Stein finds a way to explain in some manner—because a certain vagueness in the analysis of the sentient remains, not only on the linguistic level, but also on the theoretical—the double belonging of the human person to the world of nature and that of the spirit. This second aspect shows our being subject to multiple influences, so much so that the individual psyche attaches to an objective and subjective spirit. That means in turn that the mental power moves from other individuals or from a divine spirit, acting as the source of all the inner life of the human person—as stated in *Natur, Freiheit und Gnade*—and as the unifying principle of the same in its outward expression.

In any case, it seems useful to emphasize how this concept of lifepower appears in *Einführung in die Philosophie* with primary reference to the person as a living being. There, speaking precisely of the living being, Stein underlines the fact that there is a condition in it that testifies to a growth and decrease of vital force present in the organism. This is also the basis for the emergence of certain sensate dispositions which then turn outward.

The consideration of external influences on the lifepower induces Stein to widen the scope of the analysis to the community level and thus to exit from the individual sphere, to open it to the relationship with the surrounding world and specifically with that constituted by other subjects.

7.5 The Personal Dimension of the Human Being: the Core[17]

We now come to the question of the core, the second element of novelty within the Steinian conception of the living body. This term already makes its first appearance in *On the Problem of Empathy*, albeit marginally, where Stein states: "a single action and also a single bodily expression, such as a look or a laugh, can give me a glimpse into the kernel of the person [...]. Within its (of the soul) individual form we strike an unchangeable kernel, the personal structure" (1989, pp. 109–110). Therefore, there is something unchangeable that defines the individual and that expresses his or her personal structure. More, we are not allowed to know. It is in *Philosophy of Psychology and the Humanities* that the young phenomenologist returns to speak with greater specification of what is to be understood by the "core" of the person and in what context it is to be understood:

> the human personality considered as a whole shows itself to us as a unity of qualitative distinctiveness that is fashioned out of core, a formative root. It takes shape in soul, body, and mind; but only in your soul does the individuality achieve clear and unambiguous expression (Stein, 2000, p. 238).

In this sense, what most properly distinguishes us is the capacity for an inner life that is expressed and modified, and also transforms our very body; in fact, in Stein's words, "does not flow in itself, but acts, ends in an 'expression', imprinting its imprint on the body. The expression or, more clearly, what is to be expressed is a

---

[17] The term used by Stein in German is *Kern*. We take here the most widely used translation, "core", observing that in a quoted text from *On the Problem of Empathy* another term appears: "kernel", but with the same meaning of core. This concept has not been further developed by Steinian scholars, nor does it appear in the *Edith Stein-Lexikon*, unlike the concept of lifepower.

certain external condition of the body ('traits' of the face, sound of the voice and such) and particular changes that it shows" (Stein, 2010, p. 93). Therefore, even in the dimension of interiority we find an element that, in addition to being properly human and therefore common to all, characterizes us in an individual way by identifying us specifically. If, in fact, each of us already has his or her own physiognomy—linked to the external aspects of the face—it is true that these elements take on a further specific and unique connotation in the way in which each individual expresses his/her own interiority with them. The dimension of character appears here, as a spiritual dimension that gives space to something of each of us: the sphere of affective and volitional life. "Character is the ability to feel and the impulse with which this feeling is transformed into will and action [...]. Character is openness to the sphere of values and the way in which it mobilizes itself for their realization" (Stein, 2010, p. 97). Elsewhere, Stein refers these aspects more specifically to the core of the person which then expresses itself in character, giving an imprint also to the external appearance of the living body (Stein, 2010, p. 136). In the life of feeling and will, the depths of the self awaken and the soul opens up to the world of values; the inner life and character are imbued with the dormant qualities of the soul. The way values are accepted and the way we behave towards them depends on the characteristics of the soul. It seems to be something mysterious—and this is what Stein herself calls it, when she tries to find an explanation, taking up the analysis of Conrad-Martius—that the human being lives from his/her soul which is the center of his/her being, but which does not cover the totality of the ego (represented by soul, living body and spirit). Up to this point, there is nothing new with respect to the Husserlian analyses of the soul, which is defined as: "Unity of a soul life, a stream of consciousness, I, etc., a unity extending through time (the same time in which the body lasts), and it 'acts' into the physis and experiences effects from there" (Husserl, 2021, p. 540).[18] But this soul emerges for Stein from a root or core, "that determines the being of the individual soul-essence as a whole, in all its dimensions" (Stein, 2000, p. 228), as a forming principle from which the individual's being takes form. However, Stein immediately adds: "not all bodily and not all sentient beings and occurrences are 'core-valent', fashioned out of the core" (Stein, 2000, p. 228). The life of the soul, when it emerges from itself and makes its appearance before the world, is a spiritual actuality and its purity is never affected, in the sense that its qualities can neither be inculcated nor eliminated; a possible transformation can be considered as a conversion due to a supernatural force.

Looking at a passage of Husserl on the soul, a characterization seems to emerge that could correspond *in nuce* to what Stein goes on to explicate and develop with an autonomous conceptualization as the core of the soul and the person:

> The flow of the life of the soul has its unity in itself, and if the "soul" belonging to a body stands in a functional connection of mutual dependence to the material body, then the soul certainly has its permanent psychic qualities, which are expressions for certain regulated

[18] "Einheit eines Seelenlebens, ein Bewusstseinsstrom, Ich etc., eine durch die Zeit sich erstreckende Einheit (dieselbe Zeit, in welcher der Leib dauert), und sie ‚wirkt'in die Physis hinein und erfährt von da aus Wirkungen".

> dependencies in the occurrence of the psychic in its dependence on the corporeal (Husserl, 2025, p. 542).[19]

The Steinian reflection on the soul and the core presents a certain vagueness in terminology and definition: the individual soul is not the spirit, and yet its "emergence" before the world is a spiritual actuality; ultimately, "your soul's being is like the core in which it roots, an individual as such, something indissoluble and unnameable" (Stein, 2000, p. 231). There is no growth seen as a transformation of its qualities in the core as in the soul, but only a development by accretion. The soul, in fact, does not present itself from the beginning of the sensate development of an individual, but becomes gradually visible and, even if hidden, continues to transform itself and mature. When it is awakened by external circumstances, all its wealth pours into the life of the present moment; at this moment, one must not escape from one's soul because then the forces that flow from it would be lost (Stein, 2000, pp. 234–235). If the soul bursts into one's present life, sensate development benefits and if its forces are annihilated by the world to which it turns, it simply goes back into hiding. Unlike the sentient, in fact, the world of the soul has its own forces and does not need lifepower, but if its internal forces are utilized and absorbed by the external world, they can no longer be formed anew or replaced by others—the soul simply withdraws until it disappears (Stein, 2000, pp. 237–238).

Within the limits of what Stein calls "original predisposition", I can give space to certain inclinations and exercise more virtues or give myself over to vice; on the other hand, I will not be able to do what I am denied by nature. What the German scholar had already highlighted with extraordinary acumen is the very personal note that characterizes us and that will then make that score something unique and unpredictable, by virtue of freedom and the external influences of the surrounding world. In human character, there is therefore a unique and irreducible qualitative element which is the essence of the person and which unfolds in character development depending on these influences, which may be favourable or not. The permanent qualities of a person and his/her original character are situated in the depths of the soul; here the predisposition to the world of values is founded, and the center of radiation of creativity is "inserted" in the core of the person and with this, character in its most specific sense, while the sensitive abilities, memory or reason, do not depend on the core. In *Einführung in die Philosophie*, Stein compares the core of the living being in general to the pure Ego, affirming that the first is the center of irradiation of personal life as the pure self is of every experience of consciousness (Stein, 2010, pp. 131–49) and goes on to affirm that the ego cannot coincide with the core, but only find a place in it.

Now, if what has been said above mainly expresses the movement of development of the core of the person or soul outwards, in welcoming the world of values and

[19] "Der Fluss des Seelenlebens hat seine Einheit in sich, und wenn die zu einem Leib gehörige ‚Seele'in funktionellem Zusammenhang wechselseitiger Abhängigkeit steht zu dem dinglichen Leib, so hat die Seele freilich ihre bleibenden seelischen Eigenschaften, die Ausdrücke sind für gewisse geregelte Abhängigkeiten im Auftreten von Seelischem in seiner Abhängigkeit vom Leiblichen".

all the impressions that come from it, it must be said that there is also a movement entirely within the core itself—that is, the soul is also filled with something that is proper to it. "It has"—says Stein—"a depth,"; so, for example, I can decide to give more or less space to a certain experience that emerges in me, and do so randomly or intentionally, but in a form limited to a certain circumstance, or I can make that experience a characteristic feature of my lifestyle.

Therefore, the analysis of individual personality shows, according to Stein, the essence of the person not only constituted by typological traits, but also by an individual core that gives an entirely individual note to every typical aspect. "The whole world with which an individual is concerned bears the imprint of his/her personality, of his/her typical traits and of his/her personal characteristics"[20] (Stein, 2010, p. 144).

7.6 Conclusions

At the end of these reflections, we can conclude that the Steinian project of an ontology of the person is actually rooted in the first years of her work alongside Husserl. It reflects, from the time of her doctoral work and still in her works of the '20 s, a strong consonance with the development of the phenomenology of the master with regard to the living body and the conformation of the person in body, soul and spirit. Alongside the shared foundation which sees the same approach and development from the body to the psyche to the spirit passing through the intersubjective relationship, certain elements of originality emerge with which Stein expands the analysis of the Master in her own way. This is done specifically first through in-depth research on empathy and then with the introduction—probably not only for an independent understanding of phenomenological analysis, but also for personal events and sensitivities that led Stein to focus more on the dimension of the real self—of the terms that we focused on in this work: lifepower and core.

These terms are not only foreign to the Husserlian research of *Ideas II*, but show the autonomy to which the vivacity and richness of his method could lead, inspiring each person to continuous, new openings in their encounter with "the things themselves". As Baseheart affirms: "while Husserlian terminology and method are major influences, she [Stein] employs them in an original manner, differing from him in developing her own theory" (1997, p. 48). This is exactly what Stein seems to do in the face of the emergence of new personal experiences that, not being able to remain unheard in as much as they are consciously welcomed, push her to define and enrich the nature of the human person also through them. In the end, it seems to us that we can say that these elements of novelty and detachment from the *Ideas II* do not represent a rupture, nor elements of conflict with the overall discourse, but rather indicate further openings or enlargements of the same contents, through experiences simply not yet contemplated, in a continuous bond to reality.

[20] "Die gesamte Welt, an der sich ein Individuum betätigt, trägt das Gepräge seiner Persönlichkeit: seiner typischen Züge wie seiner persönlichen Eigenart".

References

Baseheart, M. C. (1997). *Person in the world. Introduction to the philosophy of Edith Stein.* Springer.

Beckmann-Zöller, B. (2010). Einführung. In E. Stein, *Beiträge zur philosophischen Begründung der Psychologie und der Geisteswissenschaften. ESGA 6*. Herder.

Betschart, C. (2009). Was ist Lebenskraft? Edith Steins erkenntnistheoretische Prämissen in "Psychische Kausalität" (Teil 1). *Edith Stein Jahrbuch, 15*, 154–184.

Betschart, C. (2010). Was ist Lebenskraft? Edith Steins anthropologischer Beitrag in "Psychische Kausalität" (Teil 2). *Edith Stein Jahrbuch, 16*, 33–64.

Calcagno, A. (2006). Assistant and/or collaborator? Edith Stein's relationship to Edmund Husserl's *Ideen II*. In J. A. Berkman (Ed.), *Contemplating Edith Stein* (243–270). University of Notre Dame Press.

Caminada, E. (2019). *Vom Gemeingeist zum Habitus: Husserls Ideen II. Sozialphilosophische Implikationen der Phänomenologie.* Springer.

Hackermeier, M. (2008). *Einfühlung und Leiblichkeit als Voraussetzung für intersubjektive Konstitution.* Verlag Dr. Kovac.

Husserl, E. (1950). *Cartesianische Meditationen und Pariser Vorträge* (S. Strasser, Ed.). Martinus Nijhoff.

Husserl, E. (1952). *Ideen zu einer reinen Phänomenologie und phänomenologischen Philosophie* (Vol. II, M. Biemel, Ed.). Nijhoff. English edition: (1989). *Ideas Pertaining to a Pure Phenomenology and to a Phenomenological Philosophy.* Martinus Nijhoff.

Husserl, E. (2001). *Natur und Geist. Vorlesungen Sommersemester 1927* (M. Weiler, Ed.). Springer.

Husserl, E. (2008). *Die Lebenswelt. Auslegungen der vorgegebenen Welt und ihrer Konstitution. Texte aus dem Nachlass (1916–1937)* (R. Sowa, Ed.). Springer.

Husserl, E. (2025). *Ideen zu einer reinen Phänomenologie und phänomenologischen Philosophie, Hua IV/V. Zweites Buch: Phänomenologische Untersuchungen zur Konstitution und Wissenschaftstheorie sowie das Nachwort zu meine Ideen. Die Drei Teilentwurfe und Thematische Ausarbeitungen* (D. Fonfara, Ed.). Springer.

Imhof, B. W. (1987). *Edith Steins philosophische Entwicklung. Leben und Werk.* Springer.

Ingarden, R. (1962). Edith Stein on her activity as an Assistant of Edmund Husserl. *Philosophy and Phenomenological Research, 23*(2), 155–175. https://doi.org/10.2307/2104910

Knaup, M., Seubert, H., & Falkowitz, H.-B. (Eds.). (2017). *Edith Stein-Lexikon.* Herder.

Manganaro, P. (2000). L'Einfühlung nell'analisi fenomenologica di Edith Stein. Una fondazione filosofica dell'alterità personale. *Aquinas, 43*(1), 101–122.

Sawicki, M. (1997). *Body, text, and science. The Literacy of investigative practices and the phenomenology of Edith Stein.* Springer.

Stein, E. (1986). *Life of a Jewish Family 1891–1916. CWES I* (J. Koeppel, Trans.). ICS Publications.

Stein, E. (1989). *On the problem of empathy. CWES III* (Waltraut Stein, Trans.). ICS Publications.

Stein, E. (2000). *Philosophy of psychology and the humanities. CWES VII* (C. M. Baseheart & M. Sawicki, Trans.). ICS Publications.

Stein, E. (2001). *Selbstbildnis in Briefen III. Briefe an Roman Ingarden. ESGA IV*. Herder.

Stein, E. (2003). *Selbstbildnis in Briefen I. 1916 bis 1933. ESGA II*. Herder.

Stein, E. (2010). *Einführung in die Philosophie. ESGA VIII* (2nd ed.). Herder.

Varga, P. A. (2016). Edith Stein als Assistentin von Edmund Husserl: Versuch einer Bilanz im Spiegel von Husserls Verhältnis zu seinen Assistenten. Mit einem unveröffentlichten Brief Edmund Husserls über Edith Stein in Anhang. In A. Speer & S. Regh (Eds.), *"Alles Wesentliche lässt sich nicht schreiben". Leben und Denken Edith Steins im Spiegel ihres Gesamtwerkes* (pp. 111–133). Alber.

Valentina Gaudiano , PhD, is Associate Professor of Philosophical Anthropology at Sophia University Institute by Florence. She is a facilitator of philosophical dialogue and trainer for educators in philosophical dialogue.

Fellow of the Center for the History of Women Philosophers and Scientists in Paderborn and Member of the international group *New Voices* linked to the Center. Main areas of research: German phenomenology and Trinitarian ontology, with focus on feelings, love, empathy, intersubjectivity, concept of person, male and female, gender issues, trans- and post-humanism.

Chapter 8
Edith Stein and Edmund Husserl on Phenomenology. The Balance of the Steinian Reception of Transcendental Phenomenology in her *Introduction to Philosophy*

Carlos Guillermo Viaña Rubio

Abstract Our paper aims to take stock of the influence of the thought of the philosopher Edmund Husserl on the early work of his disciple Edith Stein to find out how close she came to (or how far she departed from) the central aspects of the phenomenological method within the framework of the transcendental turn in phenomenology that took place almost immediately after the publication of the *Logical Investigations* in 1900–1901, but which reached the general public in 1913 with the publication of the first volume of the *Ideas*. Our thesis is that Stein understood Husserl's phenomenology beyond the limit imposed by the closed pre-phenomenological concepts of idealism and realism. Both authors agreed in affirming that the autonomy of reality does not annul the notion of "donation of meaning" which explains, rather, how subjectivity can access it within the limit of experience. To achieve our purpose, we will analyze Stein's *Einführung in die Philosophie* (*Introduction to Philosophy*) in line with the fundamental concepts of Husserl's thought that appear critically analyzed in the text. We will divide our work into three parts: first, an analysis of idealism; second, the explanation of nature as a philosophical problem; finally, third, an analysis of subjectivity and the sciences proper to it.

Keywords Idealism · Subjectivity · *Ideas II* · Edith Stein · Empathy · Transcendental phenomenology · *Introduction to philosophy*

C. G. V. Rubio
Universidad Nacional Mayor de San Marcos, Lima, Peru

C. G. V. Rubio (✉)
Universidad Nacional Tecnológica de Lima Sur, Lima, Peru
e-mail: cvianar@unmsm.edu.pe

M. Lebech (ed.), *Stein and Husserl's Phenomenology of the Social in Ideas II*, Women in the History of Philosophy and Sciences 25,
https://doi.org/10.1007/978-3-032-17339-3_8

8.1 Introduction: Stein's Collaboration with Husserl and the Transcendental Turn

The complex human and professional relationship between Edmund Husserl and Edith Stein can be partially reconstructed through her correspondence with Roman Ingarden between August 16, 1916, and February 28, 1918, the year of her break with the Master.

After submitting her doctoral dissertation, Stein became the assistant of Husserl, who needed help in organizing the publication of his manuscripts; otherwise, according to Stein, he would never have published them because he was always so busy with his thoughts. Therefore, the young assistant immediately transcribed and revised the Master's unpublished writings. Although she encountered many difficulties elaborating on this material, Husserl offered her no help in clarifying the murkiest aspects. Nevertheless, Stein was able to identify at once the most arduous and urgent knots contained in them, for her main aim was to understand the whole process of thought in front of her.

Her first task was elaborating the second volume of *Ideas Pertaining to a Pure Phenomenology and to a Phenomenological Philosophy*—a challenging task, since the material was incomplete. Moreover, despite preparing the first two parts of this work (i.e., those concerning the constitution of material and animal nature), the Master did not seem interested in their possible publication in the *Jahrbuch* because he was busy preparing other writings. Besides, Husserl constantly entrusted Stein with new manuscripts to rework, which often involved ancient material. Although she was tired of doing so, as is evident from the letter to Ingarden of April 27, 1917, she knew very well that without her help, the Master would not have published anything more.

In February 1918, however, Stein decided to leave Husserl. The reworking of the manuscripts had become almost unbearable for her, as it did not allow her space for a work of her own. Nevertheless, her reworking of the Husserlian manuscripts was fundamental because it is thanks to Stein that the second volume of *Ideas* and *On the Phenomenology of the Consciousness of Internal Time* were published. As a result, she came into closer contact with Husserl's thought, which influenced her.[1]

---

[1] The early communion between Stein and her Master is described by González Di Pierro (2016) when he affirms: "as is also known, she transcribes both the second volume of *Ideas* and *On the Phenomenology of the Consciousness of Internal Time* and some other texts, which very soon allow her to see the scope of the Husserlian project, or, perhaps more modestly, the general idea of Husserlian phenomenology, by coinciding even, in her investigations, with her master, without having known his before the achievement of her results" (34). In a previous article, we raised a consideration that could be understood as complementary to this opinion of González Di Pierro about the subsidiary nature of the work of both philosophers: "Stein's philosophical work is incontrovertible proof of her spirit of collaboration with her Master. Stein did not merely transcribe Husserl's shorthand notes. She took it upon herself to shape them, and in doing so she defined what Husserlian phenomenology was to be for her, a phenomenology that was to nuance the transcendental and idealistic turns of *Ideas I* and that entailed a philosophy not as radically transcendental as Husserl's later writings. The fact that Stein's early works and *Ideas II* have a similar structure

8.2 The Idealism Question after Ideas I: Stein's Diagnosis and Limits

After the publication of the first volume of *Ideas*, the students of Göttingen had discussed very intensely the supposed idealist turn that the Master seemed to have taken in that work. Stein did not immediately pronounce herself on this delicate problem from a theoretical point of view. Her first reflections on the theme date back to 1917; specifically, in a letter to Ingarden dated February 3, she writes:

> Now I imagine I know pretty well what "constitution" is— but with a break from Idealism. An existing physical nature on the one hand, and a distinctly structured subjectivity on the other, seem to me to be prerequisites before an intuiting nature can constitute itself.[2] I have not yet had the chance to confess my heresy to the Master […] (Stein, 1993, p. 8).[3]

---

shows that the philosopher had her vision of how Husserl's investigations should be developed. The points of convergence and divergence between the two philosophers' works show exactly where they were similar and where they differed. All in all, Stein's collaboration can be understood as an authentic testimony documenting the cooperation between her and Husserl" (Viaña, 2022, 218–219). Unless otherwise noted, all translations are mine. Where a standard English translation of Stein's works exists, I quote that edition and list it in the references. When no English translation is available—especially for *Einführung in die Philosophie* (ESGA 8)—I translate directly from the German, consulting a Spanish edition as a guide and collating with the German text in the *Edith Stein Gesamtausgabe* (ESGA), Karmelitinnen Köln: https://www.karmelitinnen-koeln.de/edith-stein-archiv-kk/gesamtausgabe

[2] In this regard, we subscribe to the opinion of Calcagno (2006) when he says that "Stein believes that one must have an absolutely existing physical nature on the one hand and a distinctly structured subjectivity on the other hand as preconditions before an intuiting nature can constitute itself. Stein, as seen from the texts on nature in the *Introduction*, views subjectivity as fundamentally related to yet distinct from nature. In other words, the ego is not absolutely self-constituting and, hence, constitutive of the surrounding world. Again, Stein is reading from certain key assumptions found in *Ideas I*—an idealism of sorts where primacy of place is given to the pure ego as the locus of intentional consciousness. Nature, ultimately, is to be suspended (bracketed) in the phenomenological reduction. For Stein, on the other hand, nature, like certain subjective structures, is fundamental for the pure ego to become aware of its surroundings and aware of itself" (258).

[3] González Di Pierro explains the meaning of this so-called "heresy", considering the advances that, currently, we have in the most complete knowledge of Husserl's thought through the publication of his *Nachlaß*, which allows us to know diverse opinions of the philosopher on some topics that were not crystallized in his published works: "Heresies which, in our opinion, were not such, since, as Stein herself would realize, this 'break with idealism' was also sustained by Husserl, since precisely the disciple, with her sharpness, had managed to grasp that the 'proper structures' mentioned in the passage constitute an equally necessary objectivity in the process of perception of reality" (González Di Pierro, 2005, 180). A decade later, he reaffirms his opinion, complementing it this time with an allusion to the deliberately ignored difference between the Kantian conception of idealism and that of Husserl: "such 'heresies' are not, strictly speaking, such, if we think that Husserl does not deny the existence of an external reality; what he denies is a reality in itself, noumenal, similar to the Kantian conception, which is different and, as we will see towards the end, is one of the reasons why it is aberrant that still today Husserl is assimilated without further ado with Kant, not only concerning transcendental subjectivity but also concerning the notion of phenomenon" (González Di Pierro, 2016, 35).

Stein later confessed these concerns about idealism to the Master, who, while grateful for the critique, maintained that he would be willing to change his view only if she could show him its necessity. Shortly afterwards, however, in the *Beiträge zur philosophischen Begründung der Psychologie und der Geisteswissenschaften* (*Contributions toward the Philosophical Grounding of Psychology and the Human Sciences*),[4] Stein would develop the analysis of "distinctly structured subjectivity"—that is, of subjectivity as intentionality (the *noetic–noematic* correlation).

The problem of idealism was the breaking point between Husserl and his disciples. The *Logical Investigations* had given the impression of having put an end to the critical idealism of Kantian and neo-Kantian imprints, and they were therefore seen as a recovery of the great tradition of the past. Husserl's disciples agreed on the return to the things and the search for essences proposed in the *Investigations*. "All the young phenomenologists were confirmed realists" (Stein, 1986, p. 250). In contrast, the *Ideas*

> [...] included some expressions which sounded very much as though their Master wished to return to idealism. Nor could his oral interpretation dispel our misgivings. It was the beginning of that development which led Husserl to see, more and more, in what he called "transcendental Idealism" (which is not to be confused with the transcendental idealism of the Kantian schools) the actual nucleus of his philosophy (Stein, 1986, p. 250).

What the "young phenomenologists" did not accept was Husserl's conclusion regarding constitution: that the world is given to consciousness—indeed, to a plurality of consciousnesses—so that every being must, in principle, be experienceable and therefore knowable, ideally corresponding to a consciousness for which it is constituted. By "idealism," however, Stein points to what *Ideas I* can be read as implying—namely, a stronger thesis: the dependence of the world's being on a cognizing consciousness. This is precisely the line pressed by the Göttingen realist Hedwig Conrad-Martius; on our reading of Stein's interpretation, this is best understood as an issue of formulation in *Ideas I* rather than a thesis Husserl would accept as his own. This formulation gave the idea of a return to Kantism and the consequent renouncement of the object and the realist ontology with which Husserl's disciples had identified their task.

The problem of idealism was the first theoretical obstacle she faced within phenomenology. In the letter to Ingarden of April 9, 1917, she wrote that she had begun to reflect on the problem of idealism and had therefore sought to read some writings of Hedwig Conrad-Martius with which she was dissatisfied. This disappointment was that, rather than a refutation of Husserl, these writings seemed to her to be based on an erroneous interpretation of the Master's thought:

> I began today with "Idealism". Besides, Fräulein Gothe showed me some notes of Frau Martius on the question of Idealism. They were, however, not a refutation of Husserl; the main argument seems to me to be based on a misunderstanding of his expositions [...] (Stein, 1993, p. 14).

---

[4] Translated into English by Mary Catharine Baseheart and Marianne Sawicki as Stein (2000).

Stein was perfectly aware that Husserlian transcendental idealism was different from Kantian idealism. It is necessary to point out, however, that when Stein speaks about idealism, she does not intend to allude to that particular movement of thought that developed in Germany in the nineteenth century, but refers more generally to "a fundamental conviction of a personal and metaphysical nature" (Stein, 2007, p. 156) that is not the result of irrefutable investigations. According to her, this view appears in Husserl only in a few places of his work and does not concern the essential core of his thought.[5]

The problem of Husserlian idealism is never systematically analyzed by Stein, perhaps because she did not consider it as the main characteristic of the Master's speculation; in fact, she gives only a few hints of it in her writings. It is however possible to recover her position from the reading of her *Einführung in die Philosophie* (*Introduction to Philosophy*). In this work, Stein's opinion regarding idealism does not immediately jump out; it is not evident, since her main intention is not to criticize Husserl but to clarify what phenomenology is.[6]

[5] As González Di Pierro (2005) points out in this regard: "it is about the solution that she proposes and clearly perceives with respect to the accusation of idealism against Husserl by the first disciples, including Stein herself; she will overcome this objection thanks to the full knowledge of the works that, as Husserl's assistant, she will integrate, especially *Ideas II*, and, above all, thanks to the intuition of Husserl's philosophical intentions in his apparent 'return' to idealism and that respond to a philosophical plan and an intellectual itinerary that Stein is familiar with. Indeed, the central problem for Husserl, namely, that of the possibility of knowledge of natural objectivity, is never abandoned by him, and this is something that his pupil understands perfectly, just as she understands the necessity and importance of the existence of a consciousness in the human being that is the guarantor –as an ideal condition– precisely of the objective knowledge of Nature, that is, a *consciousness* that, and this is important to emphasize, *is equally objective* and absolutely necessary so that the act of knowledge of the world can be completely fulfilled. Let us say that, for Edith Stein –as on the other hand for Husserl, although in a perhaps less evident way– the general epistemological conditions of nature are two: one, of an objective character, which is the complex of affairs as such; the other, of a subjective character, which is consciousness in the form of the correctness of judgment. The subjectivity of consciousness is not, thus, absolutized to the point of considering the transcendental subject as the basis of knowledge, in the Kantian manner, but, on the contrary, it implies that objective reality is given to us under the form of subjectivity and its structures" (75–76).

[6] In the following excerpts, Calcagno (2006) specifies that: "Stein's *Introduction to Philosophy* is to be read as more than background work for her future philosophical writings—it is to be read also as a commentary on and dialogue with Husserl's *Ideas*" (249). "The texts contained in the volume can be viewed within the framework of three principal objectives: first, a reworking of themes presented in Stein's earlier 1916 doctoral dissertation on empathy; second, preparatory notes for her *Beiträge* that were to serve as part of her projected *Habilitation*; and finally, a serious study of Husserl's *Ideas II*. Stein very much appreciated the originality and penetration of Husserl's text. Stein augmented her *Introduction* in scope over time" (252). "*Introduction to Philosophy* serves as proof of Stein's conviction that she was collaborating with Husserl. The text is structured similarly to *Ideas II* and *III* and contains many similar, almost verbatim, philosophical insights. Stein was not only copying Husserl but also trying to understand the richness and complexity of his thought. Perhaps Stein's *Introduction* could be read as her elaboration and development of *Ideas II* and *III*—a sort of running commentary through which she attempted to better understand and rethink certain elements in Husserl's thoughts as well as develop her own insights, which ultimately would both relate to and distinguish her from Husserl" (254). This allows us to expect the idea of Stein's

The starting point of Husserl and his disciples is the same: philosophy is not a science of experience, but of the essence. Such a search for essence can be called ontology, since the ontic structure is not the same for all objects of investigation but varies according to the areas that are considered. It is thus possible to distinguish between the ontology of nature and the ontology of spirit. According to Stein,

> [...] while the empirical sciences of nature investigate what kind of objects are present in nature and how they constitute it, the ontology of nature asks what a *material thing in general* is (i.e., what its essence is), what an organism in general is, etc. While history observes the process of development of peoples and of individual personalities who played an important role in them, ontology investigates what an individual personality, a people, a state, a historical event is in general (Stein, 2004, p. 9–10).[7]

The paths of the Master and the disciple are differentiated from the opening pages of the work. Stein finds a close connection between fact and essence and, therefore, between eidetic and factual sciences:

> Just as essence and fact do not exist separately from each other, but every empirical object has an essence, and every essence is the essence of objects, so also the science of essence and empirical science stand in a very particular relation of dependence (Stein, 2004, p. 10).

This is the first point of confrontation between Husserl and Stein; it had also been a point of conflict with his other disciples. In the first book of *Ideas*, Husserl argued that the eidetic sciences are, in principle, independent of any science of facts. However, the opposite is the case: there is no science of facts without eidetic knowledge, and consequently, all sciences of facts are dependent upon the eidetic sciences. Here, another question is at stake: the relation between fact and essence. Husserl had drawn a clear distinction between these two domains, placing greater emphasis on the latter.

> *At first* "essence" designated what is to be found in the very own being of an individuum as the What of an individuum. Any such What can, however, be "*put into an idea.*" Experiencing, or *intuition of something individual* can become transmuted into *eidetic seeing (ideation)* — a possibility which is itself to be understood not as empirical, but as eidetic. What is seen when that occurs is the corresponding *pure* essence, or Eidos, whether it be the highest category or a particularization thereof— down to full concretion. [...] *The essence (Eidos) is a new sort of object. Just as the datum of individual or experiencing intuition is an individual object so the datum of eidetic intuition is a pure essence* (Husserl, 1983, pp. 8–9).

This passage presents Husserl's ideation (*Wesensschau*) and does not state a fact/ essence split. It clarifies the methodological order: eidetic inquiry is autonomous, while the science of facts operates under eidetic determinations—a point Stein fully endorses.

---

philosophy as a middle way between the phenomenology of Husserl's *Ideas* and its critics of the Göttingen circle.

[7] We would like to bring up a past point about Stein's ontological background: "The ontological background of Stein's philosophy is a feature of her time in Göttingen, but she insists that only by appealing to the phenomenological ethos formulated and practised by Husserl can reliable philosophical results be achieved. Hence, to discover the person's essential structure, the influence of the Master is so clear, on which she builds a modified concept of the constitution with which she unifies the transcendental and ontological points of view in her philosophy" (Viaña, 2021, 158).

Up to this point, Stein's position seems closer to that of her Göttingen colleagues. Like them, she maintains that the Master had examined only one aspect of reality—the essential—*as if* severing any link with existence. However, for Stein this link is not alien to essence but intrinsic to it. The essential moment and the existential one are not alternatives but are intimately correlated.[8]

In Stein's *Introduction to Philosophy*, the issue of Husserl's idealism also arises in analyzing the subject–reality relation. Indeed, while phenomenology is an eidetic science, it also takes upon itself the task of examining the relations between the world of objects and consciousness. The subject has before him a world toward which he can direct his gaze; he can also reflect on himself and his knowledge of objects. At this point, Stein asks the following question: "How is it possible that consciousness is directed toward objects, objects that are outside itself, that have their own being, detached from consciousness: a transcendent being, as it is often said?" (Stein, 2004, p. 11). The solution to this question is fundamental for clarifying Stein's position vis-à-vis the Master. She recognizes the difficulty of the problem, and she also sees it at work in both Husserl and Kant. Nevertheless, Stein argues that the relations between consciousness and objects suggest that ontology and the critique of reason are intimately connected.

> To the essential differences of the objects correspond the essential differences of the structures of consciousness; and, since it is not possible to investigate consciousness without investigating the objects towards which consciousness is directed, we see that the ontological differences –the elucidation of the various structures of the objects– provide the thread for ascertaining the corresponding structures of consciousness (Stein, 2004, p. 12).

An essential description of consciousness is only possible if one proceeds hand in hand with the analysis of the structure of the world of objects; to the essential differences of objects there correspond essential differences of the structures of consciousness, so that one cannot investigate the one without investigating the other. Therefore, ontology and the critique of reason must go hand in hand. At this point, the distance between Stein and Husserl seems evident. The Master appears

---

[8] Nevertheless, González Di Pierro (2005) offers us the following reflection on the Steinian conception of the task of philosophy, clearly in consonance with the Husserlian conception of method, in contrast with the supposed disagreement between Stein and Husserl on idealism: "For Stein, the main task of philosophy must be to try to grasp the unity that underlies the fragmentation of knowledge, which since modernity has become more and more accentuated, especially with the institution of the natural sciences with pretensions of autonomy and separateness, and their well-determined objects of study. [...] philosophy itself will gain, according to Stein, its unity and the value of this unity, approaching the Husserlian ideal of *mathesis universalis*. This is the ultimate purpose and goal of philosophy, but how to reach it? For Stein, the method is clear, at least as far as the general premise is concerned: 'the philosophical method cannot be built based on any positive science', which is evident if we have understood that philosophy is the foundation of such fields of knowledge since it examines what is obvious and accepted for them without reflection. Therefore, the philosophical method must constitute a sphere of certainty of absolute character, some knowledge, definitive and defining. It is, therefore, necessary to 'bracket' or 'suspend' not only the results of the positive sciences, but all that is known based on experiential and empirical knowledge; the residue is precisely the realm of consciousness whose main characteristic is, as Husserl teaches from Brentano onwards, intentionality" (23–24).

to hold that the spatiotemporal world—in this context—is intentionally given, and that it has a derivative (secondary) sense as a being for consciousness. Husserl by no means denies the existence of a material world. However, in *Ideas I*, he tends to focus his analysis on the relation between the cognizing subject and the object, as well as on the subject itself, and he regards the external world as a being always and only in relation to consciousness. This emphasis does not imply that the *noema* is absent; rather, the *noetic–noematic* correlation is presupposed, even if it is not thematized independently at this point.[9] For Stein, on the other hand, it is a matter of understanding how what we call nature is given to and for consciousness, and understanding what belongs explicitly to the natural world. This relation cannot be ignored because, according to phenomenology, each *noesis* corresponds to a *noema*; to each perception, a perceived being; to each will, a willed being. The world is necessary before a consciousness, so an essential description of consciousness is possible only if it goes hand in hand with a description of the structure of the world.[10] Husserl radicalized the analysis of subjectivity, his whole philosophy's primary and essential reference. Nevertheless, Stein categorically excludes Husserl from the idealism of the Berkeleyan type because the perception of which he speaks always tends towards sensible data. Moreover, she succeeds in recovering the sense of existence within the Husserlian position, as is evident from her speech in Juvisy, because she understands that he

> [...] acutely studied the contrast between activity and passivity in the whole life of the consciousness and especially in the perception of external things. [...] Intentions, which are directed towards things, are based on a changing "hyletic material", data of sensation, which are previously given for perception, which motivate the course of perception, and which motivate the intention that dominates them. The examination of this hyletic material seems to me to be of great importance for the question of idealism (Stein, 2007, p. 586).

Stein even considers that, perhaps, the resistance coming from his disciples "compelled Husserl to concentrate more and more his efforts on a convincing foundation

---

[9] Even scholars of the stature of Hanna-Barbara Gerl-Falkovitz hold this position that invisibilizes later elaborations of Husserl's thought. Gerl-Falkowitz (1991) argues that according to Stein, nature and consciousness are related but independent of each other. They do not derive from each other: "So, there are two absolute realms: Nature and Consciousness, *neither of which is derivable from the other*. At this point Edith Stein contradicts Husserl's train of thought in the *Ideas*, wherein world is always only presently *experienced world*; 'if we suppress consciousness, we suppress the world.'" (272) Reading Husserl as saying that everything, including the world, is constituted in consciousness, she sees him as an idealist in the Kantian sense of the transcendental *ego*. However, Husserl does not deny the existence of the real world, for he describes the world as something given, to which we are fundamentally related. He writes: "This Ego of intentionality is related in the cogito to its surrounding world and especially to its real surrounding world, e.g., to the things and the people it experiences. This relation is not immediately a *real relation but an intentional relation* to something real" (Husserl, 1989, 226–227).

[10] Once again, we follow Calcagno's position when he says: "Husserl is aware of this difficulty, and nature becomes something more connected to the life of the ego in *Ideas II* and in his subsequent works in that subjects find themselves as pregiven intersubjectively in the world. Husserl affirms the necessity of a pregivenness of the natural world and the community (a structure of subjectivity) as vital for understanding the theoretical and, therefore, scientific structures of human existents, the world, and other animalia" (Calcagno, 2006, 258).

of Idealism and to consider this as the core of his philosophy, which had not been the case, far from it, at the beginning" (Stein, 2007, pp. 582–583).

The solution to the Husserlian problem of idealism admitted only two points of view, intimately related to each other, which Stein understood masterfully. The first is the irrefutable relation between essence and existence; the second aspect is the correlation between consciousness and the world. The latter is not subordinated to the former, but stands in close relation to it.

In Stein, there is a balance between subject and world, consciousness and object; this leads her to a relative rejection of the position of the Master, for whom the suppression of consciousness would mean the suppression of the world. Stein summarizes Husserl's thesis as follows:

> [...] what is it that characterizes our real world, which distinguishes it from such possible worlds, and what is it that constitutes its character of reality? Evidently the circumstance that it is an *experienced* world, and certainly not only in terms of its possibility, but as a *currently* experienced world. But then reality depends on a *currently* experiencing consciousness: if we suppress consciousness, then we suppress the world (Stein, 2004, p. 75).

In other words, Stein does not accept the absolute dependence of the world on a cognizing consciousness. However, she recognizes that perception, an act of consciousness, is necessary to determine the world's existence.

8.3 Nature, Sensation, and Constitution: Stein's Critical Reading of Husserl's Analyses of Nature

Stein's position moves in a very balanced way between the analysis of subjectivity and of nature, both terms that are indispensable and essential if one wants to understand the world. Precisely because these two elements are fundamental, Stein begins *Introduction to Philosophy* with the analysis of nature, then moving on to an examination of the relation between the world and consciousness, and successively to an analysis of the structures of subjectivity. This text allows us to understand the peculiarity and originality of Stein's ideas, not only compared to Husserl but also to other phenomenologists.[11]

[11] In this regard, González Di Pierro (2016) explains: "From a more theoretical point of view, Stein's approach to the problem of nature is very illuminating. For a science of nature to be complete, according to Stein, it cannot be limited to an ontology of nature, but must be completed with a correlative epistemology; this shows, in our opinion, something very important: that the phenomenological description and method, according to the thinker, cannot be directed only to objectivity and remain in it, as, in a way, the first disciples of Göttingen, especially Hedwig Conrad-Martius, wanted it, but, as for Husserl, for Stein, the investigation of the objective phenomenon –in this case nature– has to be directed to the subjective activity, to the subjects and their cognitive operations. This means that it will have to be directed, in the last instance, not so much to subjectivity without further ado, in a Cartesian idealistic withdrawal, but to a different objectivity that arises precisely from the epistemic activity of the subject, that is, the very genesis of the so-called scientific phenomena and the specific weight of these in the entire universe of culture" (36–37).

The first part of the text is entitled *Die Probleme der Naturphilosophie* (*The Problem of the Philosophy of Nature*); its content coincides to a great extent with the Husserlian investigation present in *Ideas II*. Stein again takes up the analyses made by the Master about causality, substance, and movement, and—as we said—almost seems to follow the path of the second volume of *Ideas* that she ordered and reworked, though always critically. Nevertheless, even when she takes up the arguments and poses the questions in total coincidence with those forms of the Master, she often gives some very different solutions.

However, it would be helpful here to explain the questions raised in these paragraphs in a more precise way to understand Stein's position fully.

The thing presents itself as an enduring entity, unique and identical, independent of any experiential subject: it exists even if we do not perceive it, and when we do perceive it, we consider it as something that already existed before us (although it is only possible to affirm this from perception, which is the act in which the natural being presents itself "in the flesh", without being the appearance of something else). All perception is always from a certain point of view. The perceiving subject is a point of orientation, and from it, he looks at the space that surrounds him. Husserl also speaks about the body as "the bearer of the zero point of orientation, the bearer of the here and the now, out of which the pure Ego intuits space and the whole world of the senses" (Husserl, 1989, p. 61).

Perception is not the lowest degree of the grasping of the object, but sensation is. In the latter, there is no intentionality of any kind; the subject is passive, while the object that encounters the *ego* is something separate and autonomous that affects it. The datum of sensation is on the boundary between the subject and the object. It is situated on the edge between transcendence and immanence; however, it is precisely from the data of sensation that the originary givenness is constituted. According to Stein, an awakened self that abandons itself simply and passively to the confusion of sensations must accept them as the manifestation of a being that shows itself to it and toward which it directs its gaze. On the one hand, sensations have something subjective to them, since they belong to a subject; on the other hand, they have an objective aspect, since what they represent is independent of the subject they fit.

In the act of knowledge, the subject must not be considered to be the one who models, at his will, a formless material. In the constitution of the real world, the subject is not in any way free, because on the one hand, a particular structure of sensible material is required from which an intentional experience can be founded. On the other hand, there is the understanding of an object without the subject having the possibility of excluding it. Stein believes that:

> A rigid law dominates the *constitution* of objectivities, that is, of their constitution for consciousness. Consciousness itself, in its original form, is a continuous flow of experiencing, in which a continuously changing present-moment forms the boundary between past experiencing, which is *co*-present in a somewhat modified form, and future experiencing (Stein, 2004, p. 67).

There is thus no possibility of subjective agency in the cognitive act, even if it is still necessary to ask what it means and how such a given structure of the sensible material is determined from sensation. According to Stein, sensory data allow for a particular interpretation that is by no means a logical deduction but "something that *precedes* all logical operations" (Stein, 2004, p. 68). For Husserl, too, there is in constitution a formation resulting from a hidden synthesis that can only be revealed analytically. An object is constituted through spontaneity, the lowest degree of which is represented by grasping, which can also be a kind of recovery of a modified object.

According to Stein, the task of ontology is the discovery and analysis of the outlines of the fundamental structure of natural reality. The *a priori* of nature sought by ontology must not be confused with the laws of the natural sciences. The ontological structure of nature belongs necessarily and absolutely to nature, and it is impossible to think of cancelling it without cancelling nature itself. On the other hand, it is possible to think of neutralizing natural laws without ceasing the processes of nature.

The discovery of the structure of reality, which regulates the hidden mode and the relations between consciousness and the world, reaffirms even more Stein's position that she does not intend to give too much space to the subject with the risk of an idealistic vision, nor does she want to stop at the simple and naive realism of several of her colleagues of the Göttingen Circle. The very close relationship between schemas and the stream of consciousness is what Stein points to as the laws of the constitution. The schema operates in perception without being understood and rules without its laws being known as such. One must move away from the natural attitude and assume a reflective attitude to understand this. Husserl had alluded to schemas in *Ideas II* as "this groundwork, this corporeal ('spatial') shape, along with the filling which extends over it" (Husserl, 1989, p. 40). Stein seems to have further extended the concept of schema found in Husserl into the structure of reality that regulates and, in part, conditions the cognitive activity of the subject.

Stein's analysis in *Introduction to Philosophy* proceeds by degrees and continuous intuitions. After having arrived at ontology, whose main task is the analysis of schemas, a question of capital importance arises that allows us to understand her position vis-à-vis the Master. In § 5, *Die metaphysische Bedeutung des Wahrnehmungsproblems; die Idealismusfrage* (*The metaphysical significance of the problem of perception; the question of idealism*), the question arises about whether

> the objects of our experience possess an existence independent of all consciousness (to express it in other words: about whether we have a knowledge of "things in themselves"), or about whether the existence of objects and their structure are conditioned by the existence and structure of a consciousness (Stein, 2004, p. 69).

Again, we are faced with the central problem of idealism that Stein poses with great theoretical clarity.

According to Stein, it is possible to give different answers to the problem posed. For example, suppose one observes the sensible data that form our genuine experience as belonging to subjectivity when one opts for the idealist point of view. If, on the contrary, they are considered as coming from outside, then one opts for naturalism. Stein polemicizes with these two positions and considers that neither is completely viable because both present contradictions and neglect one or the other aspect of reality.

Naturalists have postulated nature as an absolute and unique being and have failed to elucidate the necessary connection that binds consciousness to physical beings; they did not consider this connection in the sense that "There is no possibility of showing that something could be otherwise than what experience teaches" (Stein, 2004, p. 85), but have held that this correlation exists in general, as a necessity—the very necessity they would have wanted to deny. Necessity is not experienced, but examined, and so there is a conviction that this necessary relation is independent of experience and therefore not based on it.

To clarify the idealist point of view, Stein will refer to Berkeley, for whom things are nothing else than ideas of the human spirit; ideas are not only dependent on the subject but are to be found in the subject. On the other hand, though, the English philosopher does not hold that there is any other being beyond the spirit, whether human or divine. Stein omits this kind of idealism because there is no point in arguing against it; insofar as it does not make an adequate distinction between immanent and transcendent being, between perception and perceived being, because spirit, for such a philosophical position, is the only possible reality.

The most intense discussion is in relation to Kant, and the issues she touches upon are the examination of the doctrine of the thing in itself and the consideration of the categories as functions of thought, problems that are intimately related and that, according to Stein, represent the most significant theoretical obstacles of critical philosophy.

The doctrine of the thing-in-itself is, for Stein and for Husserl himself, the metaphysical presupposition of Kant's thought. This doctrine lacks the explicit understanding that represents one of the essential aspects of phenomenology: that there is no possibility that behind the grasped being, there is another—since the subject has in front of him a world that he can know completely, of which it is possible to grasp and trace the essential structure, for which it is not possible to think of an in itself of the soul, of history or the same eidetic or ideal objectivities separated from their experiential correlates. Stein tries to find a concordance between her doctrine and Kantian doctrine, since, for both Kant and herself, the sensible matter is beyond the subject; it transcends it, while the categories are elements that constitute a schema. The analogies between Stein and Kant stop at this point, because for her, the schemas are considered as

> [...] fundamental structures of objectivity and, at the same time, as the laws that dominate the constitution of objects. [...] We could affirm that such schemes find application only where there is an experiencing consciousness, but their being does not depend on it (Stein, 2004, p. 71).

The schema of the thing is not a real thing, but is that which transforms everything real into a thing, although the complexity of affairs is never exhausted in what corresponds to the schema.

Stein makes two criticisms of Kant. The first is that he was obstinate in wanting to affirm the thing-in-itself; the second is that he made the theory of knowledge lean on the side of the subject. Although Kant placed an objective deduction alongside the subjective deduction of the categories as the condition of the objects of experience, at a certain point,

> the subjective deduction finally appears as the one that provides the ultimate foundation, since the unity of consciousness is presented as a presupposition for the unity of the object or for the possibility of an object in general. But this linking of the two deductions is not necessary. The objective deduction can be detached from the subjective deduction, and with it comes the possibility of a different conception of the categories (Stein, 2004, p. 76).

Moreover, this different way of comprehending the categories (which, by Stein's interpretation of Kant, in any case are forms of the possibility of objective knowledge, forms of the thinkability of the real), expresses—to quote Husserl—"*with eidetic universality that which must belong, 'a priori' and 'synthetically,' to an individual object within the extension of the region*" (Husserl, 1983, p. 31). They are not contained in the subject; they are not the contents of consciousness, insofar as consciousness is not a vessel in which experiences are collected; but it is precisely the experiences that, passing one to the other, constitute the flow of consciousness. Categories are not in our spirit, nor do they have a real existence, because, according to Stein, they are real "only in the individual objects, but our attentive gaze enhances some factors of them and 'hypostatizes' them by turning them into independent essences" (Stein, 2004, p. 91). Formal categories are forms in which every being presents itself as knowable. Although they are forms of the condition of possibility of knowledge and its object, they are not functions of the subject. They do not have real existence either, but are the presupposition behind all objective existence.

However, the question of the priority between consciousness and reality remains open, and, precisely on this issue, Stein raises two questions of great importance. The first is whether a consciousness that does not have a nature in front of it is conceivable; the second is whether a nature to which some consciousness does not correspond is possible. To the first question, she answers with a renewed absoluteness of consciousness, insofar as it is possible to think of a consciousness to which the sensible material is completely lacking because it, according to its essence, does not depend on nature. Nevertheless, it is still a possibility and, as such, is thought.

To answer the second question, Stein takes up an affirmation of Husserl's present in *Ideas*; she considers this question from the idealist point of view:

> Every being must necessarily be experienceable or knowable, that is to say, to every being there corresponds *idealiter* a consciousness, for which it is constituted. *Idealiter*, this does not yet mean that there must exist an actual consciousness that precisely realizes the corresponding knowledge or experience. Generally speaking, the possibility that such a consciousness exists is enough (Stein, 2004, p. 75).

On the other hand, the idealist position maintains that being is dependent on a possible consciousness; this does not apply to mathematical axioms, since those do not depend on the subject that thinks them, but must be demonstrated or tested constantly in the same way or permanently from the same perspective. The real world is bound to be experienced, so Husserl would be right in arguing that the suppression of consciousness would entail the suppression of the world.[12] Stein does not accept this view because she is convinced that a real being is not only real as experienced by consciousness, but also independent of it.[13] While it cannot refer to a being outside of all experience (since nothing is known of a non-experiential being that cannot be checked against experience), it can indeed be said that all experience, according to its sense, refers to something independent of it.

The problem of the autonomous existence of reality concerning consciousness can be solved by a deeper analysis of the data of sensation, which we have discussed above. Even if we do not accept the results of sensualism, which considers the data of sensation as the only reality, Stein argues that this theory is correct in considering the data of sensation as fundamental for the constitution of reality. The experience of the external world is not the experience of the solitary subject; the objectivity of this experience lies in its intersubjective commonality—that different subjects can experience the same things, even if from their own standpoints—not in any independence of experience from consciousness. This allows her to say that experienced reality is, in its being, utterly independent of consciousness. Furthermore, only when it is understood that there are independent objects of experience (which are "just there") is it possible to say that it derives from the particular structure of nature to be relative to consciousness. Idealist philosophy has not made this sufficiently clear.

---

[12] We have previously presented our position on the dilemma involved in clarifying Husserlian idealism in this manner: "it is not probable that we will be able to justify the supra-empirical elements of knowledge from the acts of the subject considered in their pure facticity. For Husserl, a transcendental subject reduced to the complex of aprioristic elements present in knowing would be contradictory, since, on the one hand, the individual would be the support of the pure subject, while, on the other hand, the individual is only objectively justified from the transcendental subject. That is why, from the simplest of ontologies, we should rule out the possibility of any supra-empirical absolute reality, unless we appeal to God to grant it the condition of objectivity. But if we renounce to make theological extrapolations, it will have to be in the subject where we find the laws that legitimize all knowledge. At this point, we return to the problem that makes it so difficult to clarify the phenomenological sense of subjectivity: the laws that are fulfilled in the subject go beyond it. Methodologically, we will have to overcome pure individuality and open ourselves to the metaphysics of categorical and specific levels of being. Thus, we will be able to justify these laws which, working in the empirical subject, go beyond their pure empirical nature. However, these laws should not be sought only in the knowing subject, but also in the sphere corresponding to the known object" (Viaña, 2020, 138–139).

[13] At this point, after having dealt at several moments of our work with the theme of the suppression of the world as a necessary consequence of the suppression of consciousness, it is worth reflecting on the following quote from González Di Pierro (2005), which clarifies the difference between world and nature in Stein: "Edith Stein makes a clear distinction between world and nature: the latter is constituted by objects whose main characteristic, for her, is a certain autonomy. The world, on the contrary, is mainly made up of objects that receive the sense of their being from something that is foreign to them. Their teleological sense is to be used; the annulment of their end represents the annulment of their being. However, all objects, whether natural or cultural –and here Edith Stein precisely uses the term *cultural*– find their common denominator in their belonging, as parts of a totality, to the same spatial world. What happens is that objects possessing an end outside themselves are, in a certain sense, natural objects, but they find themselves transformed and reworked in an arbitrary way. It is interesting to note that for Edith Stein the instance through which the above distinction is filtered is precisely the human being, not only because he is himself a natural object, but because from him nature begins to transform itself" (77–78).

Stein's position on whether the object of consciousness depends on consciousness is articulated from a phenomenological standpoint: consciousness is always directed toward something other than itself—an intentional correlate whose being does not depend on the act. According to Husserl's phenomenological terminology, every intentional experience has two inseparable moments, one subjective and the other objective, which Stein calls *noesis* and *noema*. By *noema*, we do not mean a copy of the object; rather, we mean the object's sense or mode of givenness to consciousness. As sense, it is distinguishable from the thing itself, yet it is the very same object as presented. This distinction does not make Stein a representationalist or an idealist; consciousness grasps the things themselves in and through their noematic sense, and in fulfilling acts the selfsame thing can be given as identical across changing profiles. By contrast, Conrad-Martius argues for an absolute coincidence between reality and the noematic moment and, on that basis, charges Husserl with idealism. Stein counters this stance: she rejects that identification, preserves the independence of the world's being, and distinguishes the thing itself from its noematic sense.

To better explain Stein's position on this proposal, it is helpful to refer to what she discusses in § 15 of the text *Introduction to Philosophy* entitled *Der Stoff der Naturerkenntnis und seine erkenntnistheoretisch-metaphysische Bedeutung* (*Question of the knowledge of nature and its metaphysical-gnoseological significance*), since after a step by step analysis of the gnoseological problem, she asks whether or not the nature that is experienced has an absolute being. As we have mentioned, the categories (both formal and material) guarantee the possibility of knowledge of nature. At the same time, its facticity, its real existence, is assured by the presence of sensible nature. However, of a sensible material nature that does not fit into that system of forms that are not subjective but captured and hypostatized by the attentive gaze of the observer, it is not possible to have an experience. On the other hand, the forms only give a nature and a possible experience, not a real one.

Nevertheless, "The 'form' of nature, which orders the sensory material and regulates the course of our experience, we have recognized as independent of the experiencing consciousness" (Stein, 2004, p. 97). To say that these forms of nature are independent of consciousness is to deny that the subject orders reality—as in Kant's account—and to affirm that matter can receive a general form precisely because it has a general essence. In Stein, "constitution" does not mean the creation or fabrication of objects by consciousness; it designates the phenomenological explication of how objects are given and acquire sense in intentional experience. To constitute is to disclose, within the noetic–noematic correlation, the object's sense as the selfsame thing across changing profiles, and to articulate its essential form. Constitution, therefore, concerns the order of givenness and sense (*Sinngebung*) and presupposes the independence of the world's being.

8.4 Subjectivity and Its Sciences: Personhood, Embodiment and Motivation

The second part of *Introduction to Philosophy* is focused on the analyses concerning the ontic structure of the person,[14] of which Stein had already spoken in part not only in her dissertation but also in the *Beiträge*.[15]

To achieve a comprehensive theory of the experience of nature, an analysis of the subject is necessary; it is crucial to understand how a person becomes a person and what relationships he establishes with himself, with others and with the world. Therefore, it is necessary to start by clarifying the concept of person; with this concept, Stein does not refer to the isolated individual, outside any relations with others and with the world, but ourselves and our fellow men. In these pages, Stein makes a meticulous investigation of the consciousness, the *psyche*, the *Körper*, and the *Leib* as it is perceived by oneself and by others; she focuses on the relationship between the will and the body; she speaks of values and motivation as fundamental laws of the spiritual life. Furthermore, it is strictly because the subject is such a complex unity that there is a whole series of sciences that are interested in it: biology, anatomy, physiology, psychophysics, psychology, history, sociology and other sciences of the spirit, at the basis of which we have phenomenology as the doctrine of pure consciousness.[16] While these analyses ground empathy in embodied givenness, in the closing section of *On the Problem of Empathy* Stein leaves open whether empathy with a non-bodily

---

[14] Another example of Gerl-Falkovitz's misrepresentation of Husserl revolves around his analysis of the person. For her, the difference between Stein and Husserl lies in the fact that Stein emphasizes the structure of the human person in her text, whereas Husserl does not: "Edith Stein proceeds independently, in that she by no means remains only with the problem of consciousness, but stretches the framework from the beginning much further: towards the *person*, her continuing interest. [...] Body, psyche, mind are themselves only phenomena in the person, which need elucidation (and have not been subjected to such elucidation by Husserl)" (Gerl-Falkowitz, 1991, 273–274). However, one of the fragments that Edith Stein grappled with while preparing the *Ideas* was the section entitled "The Constitution of the Spiritual World". This section constitutes at least a very large part of *Ideas II* and is a profound meditation on the person. Husserl writes about the personal *ego*, empathy, and the person as the subject of acts of reason and as a free being. Stein bases much of her thinking on Husserl's analyses.

[15] These arguments would later be taken up again in *Finite and Eternal Being*.

[16] Perhaps it is necessary to insist on the relationship between Husserl and Stein for the connection between the body and the personal *ego* as a fundamental topic in the philosophies of both authors. In this sense, we take up Calcagno's view when he writes: "Edith Stein connects not only body to soul but body to a proper sense of the I, the personal ego. In other words, the I cannot be an I without a body. This connection for Stein of the body to the personal ego is significant in that she frames the discussion within the context of empathy. [...] Husserl's notion of body is given a central place in *Ideas II*, for it is conceived as a point of orientation. Husserl views the body as a 'passageway' to understanding the motivations of another subject. He does not explicitly attribute to the body the deeply personal attachment that Stein does. Stein speaks of the body as enclosing interiority and as a gateway to interiority, both the psycho-spiritual and the personal. Husserl, does, however, implicitly connect the body to the uniquely personal in that every subject and therefore every body is personal" (Calcagno, 2006, 259).

person might be possible. Accordingly, she would not defend the strong thesis that there can be no I without a body, even if Husserl might.

Also, in *Introduction to Philosophy*, there is—as already seen in the *Beiträge*—the attempt to approach the so-called sciences of the spirit philosophically. Here, Stein made a first distinction between the sciences of culture—which had the task of investigating some objective cultural products (law, economics, language, etc.)—and the historical sciences, which were interested in the development of man about the evolution of cultural products. Alongside these sciences, which she calls inductive, Stein believes that some succeed in tracing the *a priori* structure within empirical spiritual forms:

> All empirical modelings of mind have an *a priori structuration* that's not to be expounded in either "individualizing" or "generalizing" (which means inductive) consideration, but that's presupposed by both and [must become the] object of new disciplines: the *a priori humanities* (Stein, 2000, p. 302).

For these sciences, it makes no sense to speak of their origin—because languages and positive law, for example, are born, but not language as such, nor law, because they are eidetic singularities.

The same discourse is taken up again, albeit in a reduced form, in *Introduction to Philosophy*. Here, Stein makes the distinction between the empirical sciences of the spirit, which are interested in empirical objects (different persons, different states or economic forms), and the *a priori* which can identify the *a priori* structure of the realm under consideration; i.e., a general structure of meaning. In this work, however, Stein is not only interested in the sciences of the spirit, but also in the natural sciences. Empirical natural sciences such as physiology, for example, are essential for the understanding of the constitution and functions of the human organism understood, not as *Körper* (this is what anatomy is about), but as *Leib*, as a living body. What interests Stein is not so much the distinction between the different sciences but the understanding of the modalities through which the empirical natural sciences that examine man operate. Physiology, for example, does not ask what a body is because it seeks only its *de facto* characteristics; the objects with which this discipline has to do are always individuals, but this does not matter, since it only seeks from an object what it has in common with others.

> How should we understand this attitude, which is common to all natural sciences, both those dealing with organic nature and those dealing with material nature? Is it a matter of the arbitrary nature of the researchers, of a special orientation of their interests, or is it something objectively grounded? In order to decide about this, we have to reflect on what individuality means (Stein, 2004, p. 201).

This is the background to the controversy, which was raging at the beginning of the century, about the possibility of extending the geometrical-mathematical method to the human sciences; Husserl's criticism of those naturalists who wanted to enclose the *psyche* within mathematical laws is to be found in Stein's *Introduction to Philosophy*. Moreover, the question of the essence of the object, of individuality, which must be examined, is raised again here—it cannot be left unanswered if one wants to understand how things stand in the various fields of investigation.

The natural sciences use space and time as the *principia individuationis* of an object; however, the individuality of the material thing is different from the individuality of the spiritual person independent of space and time—somebody could have done different things in various places and at different times, and still maintain their individuality. Therefore, the natural sciences cannot consider the spiritual person's individuality because it is endowed with that which can in no way be reduced to mathematical laws. All scientific work aims to rediscover the general laws within which the knowledge of natural reality is possible; those laws are the result of observing a regularly occurring experience, from which a hypothesis is formulated that must be confirmed in verification through an experiment. This is the Galilean method. Therefore, the common thread of all laws is the idea of regularity, which is the law of causality in material nature; while for living organisms, it is the teleological presupposition.

The spirit, however, cannot be quantified; it belongs to a qualitative singularity that escapes any general law that can be expressed through geometrical-mathematical laws. So, if individuality escapes any rule or norm, the question arises as to how the sciences of the spirit manage to grasp it:

> it seems enigmatic in what way one can take hold of individuality itself. How is it possible for the historian to trace for us the image of a historical personality, being so that for its description he has only general expressions at his disposal? Will he not necessarily have to stagnate in what is typical? (Stein, 2004, pp. 212–213).

The historical can lead us to individuality only because it has general features. In individuality, the personal note is linked from time to time to a combination of general elements, and yet it is in itself a connection of meaning, something unique. The historian offers us an individual type, a unique idea; it is not possible to understand the personality as historically given, but one must strive to portray a personality as close as possible to historical reality. "The historian, if he succeeds in his task, can only bring vividly before us an intuition of the personality that reveals its individuality" (Stein, 2004, p. 213).

Stein's analyses continue to describe the historical method as an example of the method of the empirical sciences of the mind. Here, she does not find herself—as she does in the *Beiträge*—having to philosophically endow the sciences of the spirit. However, from these results, the philosopher shows concretely how the spirit's empirical sciences should continue to understand the subject.

In the last pages, Stein returns to a fundamental topic of phenomenology: the objectivity of historical consciousness. Still, in connection with the problem of historical reality, she asks "by what right the historian can claim to discover the true historical connection" (Stein, 2004, p. 225). Here, she takes up the polemic against historicism and rebukes Simmel's position that historical consciousness does not manage to understand reality as it is given, but instead reconstructs it in multiple ways, from its categories. In other words, the historian only brings out of the actual events that which is necessary for understanding the connection of meaning that is observed from time to time. Stein is also convinced that there is a difference between a spiritual life in its original course and history as the historian represents it.

Nevertheless, the historian cannot decide what is important about an epoch and distinguish it from what is not. "The meaning *is speaking* to us from the documents and is not generated by the observer" (Stein, 2004, p. 225). Of course, it is possible to have different accounts of the same historical epoch, but this only means that the meaning-content of the events was multiple, not that each researcher can draw a different picture of the same epoch. Moreover, if Simmel's point of view were accepted, it would not be understood why the same individual life speaks to each scholar from each representation. Here we have the *Leitmotif* of phenomenology, which was at the heart of the *Logical Investigations*: the idea of absolute truth and objective knowledge. In *What is Phenomenology?*, Stein writes: "The spirit *finds* the truth, it does not *engender* it. And the truth is *eternal*; when human nature changes, when the psychic organism changes, and when the spirit of the times changes, men's opinions may change, but truth does not change" (Stein, 2007, p. 154). This is the great historical merit of Husserlian phenomenology, which Stein always fully shared.

References

Calcagno, A. (2006). Assistant and/or collaborator? Edith Stein's relationship to Edmund Husserl's *Ideen II*. In J. Berkmann (Ed.). *Contemplating Edith Stein* (pp. 243–270). University of Notre Dame Press.

Gerl-Falkowitz, H.-B. (1991). Nachwort. In E. Stein, L. Gelber, & M. Linssen (Eds.), *Einführung in die Philosophie* (pp. 265–277). Herder.

González Di Pierro, E. (2005). *De la persona a la historia. Antropología fenomenológica y filosofía de historia en Edith Stein*. Driada.

González Di Pierro, E. (2016). La fenomenología de Edith Stein como refutación del "realismo fenomenológico" del "Círculo de Gotinga". Stein y su interpretación del idealismo trascendental husserliano. In *Acta fenomenológica latinoamericana V (Actas del VI Coloquio Latinoamericano de Fenomenología)* (pp. 27–41). Pontificia Universidad Católica del Perú.

Husserl, E. (1983). Ideas pertaining to a pure phenomenology and to a phenomenological philosophy. *First book: General introduction to a pure phenomenology* (F. Kersten, Trans.). Kluwer.

Husserl, E. (1989). Ideas pertaining to a pure phenomenology and to a phenomenological philosophy. *Second book: Studies in the phenomenology of constitution* (R. Rojcewicz & A. Schuwer, Trans.). Kluwer.

Stein, E. (1986). *Life in a Jewish family: An autobiography, 1891–1916. CWES I* (J. Koeppel, Trans.). ICS Publications.

Stein, E. (1993). *Self-portrait in letters, 1916–1942. CWES V* (J. Koeppel, Trans.). ICS Publications.

Stein, E. (2000). *Philosophy of psychology and the humanities. CWES VII* (M. C. Baseheart & M. Sawicki, Trans.). ICS Publications.

Stein, E. (2004). *Einführung in die Philosophie. ESGA VIII* (C. M. Wulf, Ed.). Herder.

Stein, E. (2005). *Obras completas II: Escritos filosóficos (Etapa fenomenológica, 1915–1920* (C. Ruiz Garrido & J. L. Caballero Bono, Trans.). Ediciones El Carmen; Editorial de Espiritualidad; Editorial Monte Carmelo.

Stein, E. (2007). *Obras completas III: Escritos filosóficos (Etapa de pensamiento cristiano, 1921–1936* (A. Pérez Monroy, J. Mardomingo, & C. Ruiz Garrido, Trans.). Ediciones El Carmen; Editorial de Espiritualidad; Editorial Monte Carmelo.

Viaña, C. (2020). Edith Stein y Edmund Husserl en diálogo en torno al idealismo trascendental. In *Edith Stein una filósofa de nuestro tiempo* (pp. 137–162). Aula de Humanidades, UPAEP.

Viaña, C. (2021). Edith Stein as a middle way between the phenomenology of Husserl's *Ideen* and the critics of Göttingen circle. *Revista do NUFEN, 13*(1), 147–167.
Viaña, C. (2022). Edith Stein en su actividad como asistente de Edmund Husserl en Friburgo. *Eikasía Revista de Filosofía, 109*, 191–220. https://doi.org/10.57027/eikasia.109.334

Carlos Guillermo Viaña Rubio (Lima, Peru, 1985) is a Tenured Professor in the Department of Philosophy at the *Universidad Nacional Mayor de San Marcos (UNMSM)* and in the General Studies Program at the *Universidad Nacional Tecnológica de Lima Sur (UNTELS).* He earned his Bachelor's and Master's degrees in Philosophy, with a specialization in the History of Philosophy, from UNMSM. He is currently pursuing a PhD in Philosophy at the *Instituto de Investigaciones Filosóficas "Luis Villoro"* of the *Universidad Michoacana de San Nicolás de Hidalgo* (Morelia, Mexico), where he is writing his doctoral thesis on Edith Stein, focusing on empathy, motivation, and community. Author of several publications, he is a member of the *Círculo Peruano de Fenomenología y Hermenéutica* and the *Círculo Latinoamericano de Fenomenología.* His research centers on phenomenology, intersubjectivity, and its dialogue with Scholasticism.

Chapter 9
Stein's Contribution to the Phenomenological Foundation of the Social Sciences in Conversation with Schutz's

Mette Lebech

Abstract In contrast with Edmund Husserl and Max Scheler, Edith Stein proposes a distinct object of study for psychology, the *psyche*, implicitly thereby criticising the fusion of phenomenology and psychology which can be found in both Husserl and Scheler. In *Psychic Causality*, she conducts a constitutional analysis of the psyche as a unity constituted from all acts in which life power plays a role. Life power presents in vigour and tiredness, in the positive or negative impact of emotions, and is distinct from motivation in that it cannot be "felt into" or followed; it seems to be restored when I am at rest and spent in activity; and it can be studied, as analogous to a battery, in the economy of mental life.

Life power also manifests in our personal, institutional, and animal relations, and thereby contributes to the formation of community, society, and mass. As electricity can power several machines on the same network, so psychic energy can be shared between several persons, in a number of essentially different ways. This diversity distinguishes the different forms of sociality, such that it is key to understanding how the individual contributes to intersubjective constitution.

This chapter attempts to show how the social dimension of the psyche gives rise to the idea of social causality and argues that this is what is studied by the social sciences. In doing this, it is providing an account of the object of the social sciences that is compatible with that of the Schutzian tradition.

Keywords Stein · Schutz · Husserl · Phenomenology of the social sciences · Social causality · Psychic causality · *Philosophy of psychology and the humanities* · Lifeworld · Values · Motivation

---

M. Lebech (✉)
Maynooth University, Maynooth, Ireland
e-mail: mette.lebech@mu.ie

M. Lebech (ed.), *Stein and Husserl's Phenomenology of the Social in Ideas II*, Women in the History of Philosophy and Sciences 25,
https://doi.org/10.1007/978-3-032-17339-3_9

9.1 Introduction

While it has become quite well known that Stein contributed substantially to the phenomenological understanding of empathy (Moran, 2004; Overgaard, 2019), her contribution to the phenomenology of the social is still much less well known (Beckmann-Zöller and Gerl-Falkovitz, 2006; Calcagno, 2014; Gaudiano, forthcoming; Riess, 2010).[1] It can nevertheless be of substantial benefit to social ontology, to the theory of social acts, and to the philosophy of the social sciences in general (McGuirk, forthcoming; Szanto & Moran, 2025).

To give an account of Stein's phenomenology of the social, we shall first contextualise it in relation to Schutz's better-known account (1). Next, we shall discuss Stein's understanding of the psyche (central to her understanding of the social) and her understanding of the psychic person flowing from it (2). This equips us to then discuss the role of psychic causality in social life, as well as the agency the person has in relation to it (3).

It shall be argued that what is here referred to as "social causality"—the compound impact of psychic causality on social life, the main ingredient in what we casually refer to as social pressure or social conditioning—is the subject uniquely studied by the social sciences. This was obliquely seen by Stein, but not explicitly argued, since her focus was on the sciences of psychology and the humanities. The idea underlies, however, her work on the state (Stein, 2006), as the phenomenon that brings the state into existence and conditions its persistence.

9.2 Stein's Phenomenology of the Social in Relation to Schutz's

Schutz's investigation in *The Phenomenology of the Social World* attempts to employ Husserl's *Phenomenology of Internal Time Consciousness* and *Formal and Transcendental Logic* to phenomenologically clarify Weber's basic concepts. Stein's point of departure for clarifying the formal object of the social sciences is, in contrast, a synthesis of Scheler's social ontology with Husserl's *Ideas I–III*. Schutz's and Stein's versions of "Husserl" are thus slightly different: Stein's "Husserl" is the early Husserl, who she interprets through Reinach's understanding of the social act. Schutz's "Husserl", in contrast, is the early Husserl of the *Phenomenology of Internal Time Consciousness*, as well as the later Husserl, whose "genetic phenomenology" of *Formal and Transcendental Logic* explicitly developed the complementarity of eidetic and constitutional phenomenology. It is not very well understood—and Schutz may not himself have known—that Stein, alone among the Reinach students of

[1] The present article is an expanded version of one translated into Portuguese, to be published by EDIPURCS (full reference forthcoming). It builds on a series of presentations given in Porto Alegre to the sixth Edith Stein Symposium November 2023, which in turn developed the original presentation given at the conference, of which the current volume presents the proceedings.

the Bad Bergzabern Circle (Feldes, 2015), albeit belonging among the so-called "realist phenomenologists", insisted on the same complementarity. *On the Problem of Empathy*, for example, has Chapter 2 conducting an eidetic analysis of the act of empathy, while chapters 3 and 4 complete the analysis by discussing what it is that empathy indispensably contributes towards *constituting*; i.e., the psycho-physical individual (Chapter 3) and the person (Chapter 4) (Stein, 1989).[2] The difference in emphasis between Schutz and Stein may be exaggerated by the fact that Schutz associates Stein with the "realists" who were critical of Husserl's transcendental turn and against whom strands of the phenomenological tradition—including Schutz—turned, in solidarity with Husserl. This exaggeration, however, militates against the Schutzian tradition finding a critic and a potential ally in Stein, and against Steinians profiting from engagement with Schutz. A more productive route is indicated by Wolfgang Riess, who points to the compatibility of Steinian and Schutzian themes (Riess, 2010, pp. 385–6).

The slight difference in time and emphasis, however, and Schutz's scepticism regarding "empathy" (Schutz, 1967, p. 20), which may have been directed against Stein's *On the Problem of Empathy* (Burns and Vendrell Ferran, 2024), influences their respective understandings of how meaning and motive relate to motivation and values. Schutz considered Stein's method in *Philosophy of Psychology and the Humanities* to be "a direct application of the method of eidetic reduction to unclarified notions of common sense thinking or to equally unclarified concepts of the empirical sciences" (Schutz, 1962, p. 140). This judgement probably prevented him from reading the work, or—if he did glance through it—from understanding its value. Schutz may have thought the second treatise of *Philosophy of Psychology and the Humanities*, "Individual and Community", could be read without its predecessor, "Psychic Causality". As Hagengruber (2006, p. 64) writes, however,

> Stein refers with the causal, physiologically and functionally rooted concept of *life power* to a happening that on the one hand is realised in principle in the individual and on the other essentially reaches beyond the individual and even determines it. *Life power* is in Stein an essential element of the potential individuals have for community, as psychic causality and superindividual reality merge in it. This exchange represents a spiritual and real experience. Here lies the central pivot of Stein's argument concerning social phenomenology wherefrom she founds the cultural sciences and the humanities.[3]

---

[2] That she does not discuss the importance of the result for the constitution of the idea of empathy should not be taken to mean that she does not understand there to be any feedback mechanism from constitution to the *eidos*, but it does mean that she does not consider it possible to have empathy contributing to the constitution of something without empathy being something that can be identified.

[3] "Mit dem kausal, physiologisch und funktional verankerten Begriff der Lebenskraft fundiert Stein ein Geschehen, das einerseits grundsätzlich individuell realisiert ist, das aber andererseits wesentlich über das Individuum hinausgeht, ja dieses bestimmt. Die Lebenskraft wird bei Stein zu einem wesentlichen Faktor für die Fähigkeit der Individuen zur Gemeinschaft, denn hier gehen psychische Kausalität und überindividuelle Wirklichkeit ineinander. Dieser Austausch stellt eine *geistige* und *reale* Erfahrung dar. Hier liegt der zentrale Drehpunkt von Steins sozialphilosophischer Argumentation, aus der sie die Kultur- und Geisteswissenschaften fundiert." All translations into English from works not listed in the bibliography in their English translation are by me.

If Schutz attempted to read *Individual and Community* on its own, it makes sense that he would have formed the impression he communicates. He also considered Scheler's distinction of different forms of sociality (in his case mass, community, society and *Gesamtperson*), relying as it does on Tönnies' distinction between community and society, to remain unclarified because "the collectively determined acts [...supposedly performed by each of them...] remains unclarified" (Schutz, 1962, p. 140). Schutz rightly rejects Scheler's replacing of empathy with "inner perception" (Schutz, 1967, pp. 23–24), as does Stein (Lebech, forthcoming). But he seems to ignore the fact that Stein does account for why the forms of sociality differ. To her, sentient contagion is the main way to share life power (Betschart, 2011) in mass and thus is what characterises the mass as such. Founding acts of choice determine the circulation of life power in association, and it is therefore these that characterise association (society). Value response, in contrast, predominates as the way to share life power in community, and thus I know that I form part of a community when I share life power with its members "naively" in this way. Stein would not regard each superindividual subject as being constituted from the acts it performs (since not all forms of sociality or all social formations perform acts at all, which may be what Schutz assumed). It is the individuals, in contrast, who constitute the superindividual entities and thereby make possible acts performed by representatives of those superindividual entities. However, we as individuals are frequently not at all aware that we are forming social entities by the way in which we share life power with other individuals. And yet, we can distinguish easily enough between forms of sociality, because the way we form part of mass, society and community differs essentially: I identify them from the way I participate by sharing the life power that binds them together—through psychic contagion, choice or value response. In this way, psychic causality sculpts sociality to allow for social formations which are not open to purely spiritual persons: mass and society.

We propose to show in what follows how Stein's constitutional analyses of the different forms of sociality in fact provide the transcendental underpinning for a phenomenology of the social world, as Schutz also sees it.

Others have pointed to this function. Thus Hagengruber proposes that Stein:

> proposes a sketch of the comprehensible reality of commonalities through the phenomenological method. This reality is of the order of the with-one-another of the individuals in so far as their spiritual and psychic capacities is constitutive for it. The task consists in pointing to the *a priori-transcendental* presuppositions necessary for scientific justification within the context of the analysis of the relationship between individual and community (Hagengruber, 2006, pp. 61–2).[4]

Calcagno also proposes that:

[4] "Sie entwirft ein Bild einer durch die phänomenologische Methode erfassbaren Realität der Gemeinschaftswesen. Diese Realität ist der Bereich des Miteinander der Individuen, soweit ihr geistiges und psychisches Vermögen dafür konstitutiv ist. Die Aufgabe besteht darin, die apriorisch-transzendentalen Voraussetzungen, die notwendig zur Wissenschaftsbegründung sind, im Rahmen der Analyse des Verhältnisses von Individuum und Gemeinschaft aufzuweisen."

> what makes these experiences [of mass, association, and community] distinct, is that each of them is interiorly experienced or lived in diverse ways in consciousness. [...] Stein claims that it is the unity of the bond that exists between members of a given social relation that essentially distinguishes one form of sociality from another (Calcagno, 2014, p. 112 ff.).

A standard critique raised against Schutz has been that he did not provide this transcendental underpinning, instead remaining in the natural attitude for a significant part of his analysis (Schutz, 1967, p. 97 ff.; Hall, 1977; Peritore, 1975). The topic deserves a much longer treatment than what can be provided in the context of this article, where we shall content ourselves with pointing to where the two accounts meet, to show that—far from excluding one another—they in fact can support and reinforce each other.

One of the many advantages of discussing the compatibility between the social phenomenology of Stein and Schutz is to be able to discuss the "common sense notions" of community, society, mass, and state in the context of the empirical social sciences, while being aware of the nature of their intersubjective and transcendental constitution. It is true that this constitution relies on a fully clarified understanding of empathy, such as the one provided by Stein in *On the Problem of Empathy* and supported by Husserl in his *Cartesian Meditations* (Stein, 2014a, pp. 159–61; Husserl, 1960). These accounts have their difference in emphasis, like Schutz's and Stein's accounts of social phenomenology and the early and the later Husserl's accounts of meaning, but they each shed light on the other, provide context for each other, and *together* contribute what also Schutz needs (Schutz, 1975a).

Schutz does not refer to values in *The Phenomenology of the Social World,* even if Weber did (Riess, 2010, p. 311). Stein, in contrast—being more substantially inspired by Scheler than Schutz was—describe values as motivating powers that explain not only personal, but also individual and social behaviour and action, as well as the formation of community (Hagengruber, 2006, p. 65; Lebech, 2004, 2015). Carefully describing our experience of value in motivation, Stein distinguishes it from psychic causality, which—as a type of causality—is experienced as *influencing* experience, not motivating it. How exactly this distinction relates to Schutz's distinction between the "in-order-to" motive and the genuine because-motive (Schutz, 1967, pp. 86–96) is difficult to determine, but it may bear no relation to it at all. Underlying Schutz's analysis is a reflection on time (based on an interpretation of Husserl, 1991), which—to the extent that it excludes timeless eidetic availability—Stein would understand to compromise the intelligibility of experience as such and render phenomenology impossible. It appears Schutz's later studies of Husserl may have nuanced his position (Schutz, 1975b, pp. 108–10), even if he remains sceptical of the possibility of "transsubjective validity" (ibid, 113)—which the *a priori eidos has* through empathy, for Stein.[5] Given this, it may—pending further investigation—be most prudent to assume that Stein's distinction between motivation and psychic causality is not held

[5] By this, I mean that I cannot experience someone to experience something unless I take this something to be something that can be experienced. Thus, the something experienced must be something, and the identification of this something is made possible through the *eidos*, whether by me or by you. Your identification of this something through the *eidos* is available to me through empathy (even if imperfectly), and thus I access your access to the *eidos* (or the *eidos* through you),

by Schutz (just as it is not held by Husserl), while emphasising that this does not make Stein's investigations incompatible with the investigations of either. Husserl's understanding of the contrast between causality and motivation (Rang, 1973), which Schutz's endorses, makes up for this, without explicitly recognising the idea of *psychic* causality. Making (more) room for the phenomenon of psychic causality might add significant nuance to the phenomenology of the social, since it allows Stein to show how mass, society and community differ by virtue of the different ways in which psychic causality and motivation contribute to their constitution: whereas the mass is constituted from sentient contagion exclusively, and the community from value response exclusively, the association is constituted from acts of founding and choice. As motivated acts, these are spiritual; but as effective acts, psychic causality is effective in them.

I shall assume Stein's understanding of empathy to be compatible with but not identical to Husserl's (Calcagno, 2014; Overgaard, 2019; and Payo, in this volume), as the distinction between motivation and psychic causality and the observation of the phenomenon of psychic causality are not essential to the understanding of empathy, but are consequent on it. The importance of the former distinction and of the latter phenomenon is argued by Stein in *Philosophy of Psychology and the Humanities,* likely because she could not convince Husserl to let her revise the manuscripts of *Ideas II* to take account of them to her satisfaction (Stein, 2014b, letter 1, p. 27). The contribution which *Philosophy of Psychology and the Humanities*[6] makes is thus a contribution to the project of the *Ideas*, which as a project is conceived by Stein to be incomplete in this regard (Lebech, 2023a).

9.3 Stein's Understanding of the Psyche and of the Psychic Person

In *The Crisis of the European Sciences and Transcendental Phenomenology*, Husserl identifies "pure psychology with transcendental philosophy as the science of transcendental subjectivity" (§ 72) (Husserl, 1992, p. 261). This seems to presuppose that Husserl does not envisage any subject to be non-psychical, and that therefore he identifies pure psychology with transcendental phenomenology. He also seems to take it for granted that empathy would always involve a body.

In contrast, Stein insists in the last section (IV, § 9) of *On the Problem of Empathy* that all subjects may not be psychical:

---

which in turn widens my intuition of it to the extent that it allows me to access aspects of it that I could not access if left to myself. For example, that the *eidos* is available to others explains, that we can reach linguistic understanding because of this. All of this, of course, does not compromise the first-person nature of the investigation. See also Lebech (2019).

[6] Literally: *Contributions Towards the Philosophical Justification of Psychology and the Humanities* or *Beiträge zur philosophischen Begründung der Psychologie und der Geisteswissenschaften.*

> But how is it with purely spiritual persons the idea of whom certainly contains no contradiction in itself? Is no exchange between them conceivable? [...] I have to leave the answering of this question to further investigation and satisfy myself here with a "*non liquet*" "it is not clear" (Stein, 1989, p. 118).

According to Stein, purely spiritual persons (i.e., non-psychic persons) might exist and might communicate. The examples Stein proposes are, on the one hand, the experience of grace, in which the subjectivity of God is met in faith (see also Stein, 2014a, pp. 8–72; Lebech, 2023b); and on the other, the protective spirit—for example, the *daimon* of which Socrates spoke, who indicated for him the path to take. In these two cases, the reported communication does not seem to pass by means of corporality, nor by means of the psyche, as neither God, nor angel or demon, is understood to possess body or psyche according to their essence. Transcendental phenomenology ought therefore, according to Stein, to leave at least open the possibility that there may exist non-psychic subjects—"pure" persons, as we could call them. To Husserl, in contrast, it seems all communication between subjects is mediated through their bodies. He associates the psyche primarily with the spatiality of one's body (Husserl, 2003, p. 137, § 23) and sees subjects as having soul by virtue of inductive causality passing through their bodies. The inductive nature of this causality seemingly suspends it from being experienced *in or pertaining to the stream of consciousness itself*. Stein, however, insists that there is a direct experience of causality *in experience*, which manifests its efficacity within experience (Stein, 2000, I, I, pp. 9–20). She envisages the psyche first and foremost as the mechanism of spending and restoring energy in lived experience and considers it to be constituted from the noetic contents of acts in which *life power* plays a role (Stein, 2000, p. 27). Psyche, for Stein, is in consequence an attribute of a *type* of subject, not a quality of the subject in and of itself—not all subjects have soul (i.e., a substantial unity, susceptible to psychic causality) or psyche. To Stein, the psyche is also body-bound in such a way that the experience of life power contributes to my constitution of the body, whether in myself or in others through empathy. Emotionality hence plays a significant part in my constitution of myself as a bodily subject: I feel the presence of myself throughout the body in emotional states (my heart pounds from fear; my forehead covers in sweat with nervousness; I get a bounce in my step from joy).

That Stein regards the psyche as a quality pertaining to a subject which makes it into a specific type of subject means she can make a clean distinction between the psychic subject (the psycho-physical individual) and the person (the spiritual subject of value response or motivation). The psychic person is consequently the specific type of person who is also a psycho-physical individual. The human being is such a psycho-physical person, but it is not the only possible one. Its body manifests it as of a specific type (the human type) of psycho-physical persons in the physical world *because* the subjects of my intersubjective awareness co-constitute it as the most sensible "explanation" of its and their experience. This constitution need not be uncontroversial (Lebech, 2009) and it certainly leaves open the possibility of there existing other types of psycho-physical individual persons; e.g., "Martians," the "Borg," or other aliens to the variability of which science fiction attests.

The psychic or sentient person is a person (i.e., a value valent, spiritual subject) whose experience is not only motivational—as would be the case for "pure" persons—but also psychical. Among other things, that means it experiences motivating powers, i.e., values (on which we will focus here—Stein addresses the questions regarding sensory perception elsewhere), in two ways: through value-feeling, which is a direct intuition of the motivating power, and through the effect the motivating power exercises in the psyche felt as emotion (see Lebech, 2017). That it is easy for us to make this distinction, even if we frequently overlook it, is evident from the fact that we can be unhappy about a feeling we have because we regard it as inadequate to what we experience. I may experience myself as being unable to grieve in the way I think would be indicated by the situation I am in, or lack the ability to feel happy for someone, despite thinking I ought to. This testifies to the composition of my experience: my sentient emotionality may or may not adequately reflect what my spiritual person grasps. The experience of value is stretched between a spiritual and a psychic component. When I am busy or tired, overstretched or bored, excited, nervous, or depressed, my experience of value is "affected" in its emotional component. I may well still be able to assess the value directly in its motivating power and observe my own emotion and from it learn that I am excited, nervous, or overstretched. But as these states are experienced as causal in nature, I cannot at will make them go away, even if I know they prevent me (for example) from fully appreciating the kindness of a friend's concern.

That the psychic person is a person, however, means that it is subject to motivation (Stein, 2000, p. 41). Motivation is neither blending (of simultaneously or sequentially ebbing phases of experiences), nor association (tying together of experiences). Both blending and association are phenomena that involve psychic causality, and as such they are not spontaneous, but passive, processes that we undergo (Stein, 2000, pp. 9–15). In contrast, motivation is fulfilling *activity*. It is explained by motivating powers, values, which make the motivation intelligible. The (intersubjective) intelligibility of the motivation is what allows us to talk about the value as the "motivating power" that explains the motivation, to constitute it as such and to discuss its nature and importance relative to other motivating powers (for the purposes of collaboration, for example). Stein refers to motivatedness in its entirety as "spirit", in the sense in which the Humanities are sciences of the spirit (*Geisteswissenschaften*). The person is thus by nature or definition spiritual, in so far as it is identified as the subject of value response. When the psychophysical individual is also a person, this person's acts are marked by life power or psychic causality: emotions, for example, testify to us being "impure" persons, persons who are also psychical. They even testify to our specific characteristics as psycho-physical individuals, which may vary according to the environment, health, gender, and other individual traits or circumstances. Psychic causality, in contrast with physical causality, cannot be reliably measured, since it is identified *in* essentially motivated acts (it is essentially experiential). It can, however, be transferred through emotional contagion, and be deliberately deployed through choice, and thus it contributes to the formation of superindividual entities. In this way, the psychic person is by nature (i.e., as such) exposed to both psychic and social causality. It is because human beings are psychic beings that both psychic and social mechanisms as well as social conditioning can affect them. Social mechanisms and social conditioning do not affect "pure persons", since they—according to their essence—have no causal, no psychic, dimension.

9.4 The Role of Psychic Causality in Social Life

Social causality—the compound effect of psychic causality in the collective, intersubjectively constituted, identities in which life power plays a role—plays, like psychic causality, an ambiguous role: it gives support to the individual and exposes it to manipulation; it unifies a group and divides it; it facilitates work in common as well as social exclusion; it makes solidarity possible and creates scapegoats. Like how psychic causality makes psychic mechanisms possible (e.g., projection, introjection, splitting), social causality makes social mechanisms possible (e.g., self-fulfilling prophesies, pressure guiding the asocial person to power, scapegoating). Social pressure "works" because the human being is sentient and can be caused in a dimension of its experiencing; social causality is effective because psychic causality is. Like psychic causality, social causality is therefore in itself neither good nor bad, but it testifies to the finitude of the psychic person, as well as to its dependence on life power (energy), because psychic causality does. As a result, social causality is a factor in all social formations of which psychic persons may form part. Hagengruber (2006, p. 67) describes it as follows:

> Determining the connection of individual and community through "life power" as a unity of formal elements and content defines the specific view Stein takes of the foundation of psychology and the humanities. The innovation consists in deriving this foundation through a philosophy of social construction [sozialphilosophischen Konstruktion]. In this her foundation strategy differs from Husserl, who in his *Philosophy as Pure Science* saw the foundation of the sciences to rely on an epistemologically motivated and methodologically active phenomenology. Stein takes over Husserl's methodological presuppositions, but she deploys them in an architecture of thought through which issues a socially constructed justification of science. Stein's definition of community is the condition for a possible justification of science. Marianne Sawicki (2003) sees in this the great difference to Husserl's constitution of the Social.[7]

In so far as Hagengruber is right, the social constructivist followers of Schutz—Berger, Luckmann and their students—can all recognise ancestry in Stein, since her

[7] „Den Zusammenhang von Individuum und Gemeinschaft auf der Basis der ‚Lebenskraft' als eine Einheit formaler und inhaltlicher Elemente zu bestimmen, definiert die besondere Perspektive, die Stein im Hinblick auf die Begründung der Psychologie und der Geisteswissenschaften einnimmt. Die Novität liegt darin, diese Begründung aus einer sozialphilosophischen Konstruktion abzuleiten. Insofern unterscheidet sich ihre Begründungsstrategie von Husserl, der in seiner Philosophie als strenge Wissenschaft die Begründung der Wissenschaften auf der Basis der erkenntnistheoretisch motivierten und methodisch wirksamen Phänomenologie anvisierte. Stein übernimmt Husserls methodische Prämissen. Sie führen sie jedoch zu einer Architektur des Denkgebäudes, aus der eine sozial konstituierte Wissenschaftsbegründung hervorgeht. Steins Gemeinschaftsdefinition ist die Bedingung einer möglichen Wissenschaftsbegründung. Marianne Sawicki (2003) sieht darin den großen Unterschied zur Husserlschen Konstitution des Sozialen."

view would have informed her editing of *Ideas II*. The forms of sociality are socially constructed in the sense that they are essentially intersubjectively constituted. They are socially constructed in characteristically different ways: The mass relies for its constitution exclusively on sentient contagion bringing individuals together in the experience of a superindividual stream of experience. When it comes to forming part of a mass, neither choice nor value response is relevant, only the sentient energy transmitted through the network of psyches linked by proximity (Stein, 2000, pp. 241–54). Calcagno formulates it in the following way: "Stein describes the unity of the bond between members of a mass as 'psychic contagion' [...as...] involuntary or instinctual imitation" (Calcagno, 2014, p. 113). An association, in contrast, is constituted—socially constructed—from acts of choice or institution. I chose to associate with this person or be a member of that association, and the predominating other members in turn chose to appoint me or accept me as part of their association, for reasons that have to do with objective aspects of my personality or reality (Stein, 2000, pp. 255–60). Calcagno sums it up as follows: "Societies are founded, – they require a founding act; they do not naturally grow out of human relationships" (Calcagno, 2014, p. 115). In Stein's own words: "Just as a claim arises through an act of promising and through a waiver is removed again from the world, so is the association created and annihilated through spontaneous acts of institution and disbanding" (Stein, 2000, p. 260). Thus, I do not become part of an association by mere sentient contagion, nor by personal value response. The superindividual entity of which I experience myself as forming a part is, in other words, not an association or institution if I am not choosing to form part of it, since in that case it cannot be constituted for me as such. The fact that I am motivated by the values that an association was founded to realise makes me at most a potential member of the association. In fact, I do not *have* to share the values espoused by the association if the formal procedures for adopting me as a member have been accomplished or if I have been appointed to this or that position, fulfilling this or that functionality. But I do have to accept these acts of institution as establishing a matter of fact around a goal set by the association. If I took them for nothing, the association would not be constituted for me. Acceptance, however, relies for its intersubjective credibility on the community, and thus "the only essential fact for the association is that it presupposes some mode or other of community." (Stein, 2000, p. 261). The community may exist without society, but associations are dependent on community. Society shapes community, but can neither invent it nor replace it.

In contrast to the two former types of sociality, which are constituted from sentient contagion and the deployment of psychic causality through acts of choice in relation to a preexisting community, I become a member of a community by growing into it with my whole personality; coming to reflect it as a typical member in so far as I (consciously or subconsciously) espouse its values, those that motivate it according to its essence, and by being joined by others who are similarly motivated (Stein, 2000, pp. 261–294). "[O]ne lives in and with the other" in solidarity in the community (Calcagno, 2014, p. 119). The community is constituted from the noetic components of the specifically personal acts in which the persons express themselves in their value-valence. It is thus spiritual in nature. In it, psychic causality plays only the non-essential role of strengthening the bonds between the members of the community, who can—without overruling their personal value response—remain open to influences stemming from each other because they share fundamental values.

The psycho-physical person is thus formed by both psychic and social causality, but it preserves its personal agency in relation to the latter because it has personal agency in relation to the former. This agency is best exercised when there is understanding of the nature of psycho-social causality and its rootedness in self-constitution. My agency, in fact, depends on how I obtain and manage the life power I source in various ways, because this determines my availability for the different types of social formations. To the extent that I get my energy from the physical world, I get physical energy for whichever project I might have, no matter the details. To the extent that I receive my psychic energy from others, my value response is directed through my reception by their response (in the mass, this happens directly; in associations, through the intermediary of decisions). If I let myself be motivated directly by the values and thus receive my energy directly from the motivating powers, I am available for the co-motivatedness which is constitutive for community. In this case, I retain the power of preference, but also the task of preferring. I am free in relation to all the types of common experience by precisely the same means: I can prefer values. In them, I find the motivating powers to be directly motivated, and thus I can steer the mass by value response rather than being directed by its mood. In values, I find reasons for choices, so I can make informed decisions about which associations I form part of, and why. It is thus my personal value-valence that is my agency in relation to social causality: associating with others to accomplish change through the deliberate use of social causality is the institutional possibility opened by acts of institution and choice.

It is significant that Stein conceives of grace as a personal encounter that both suspends me from the natural realm (inclusive of its socially constructed reality), liberates me, and enlivens me. Her treatise *Freedom and Grace* (Stein, 2014a, pp. 8–72), written in prolongation of her work on social phenomenology, makes use of this work, to the point of the treatise being inconceivable without it. She describes it as liberating by dispensing me from being dependent on the natural economy of the psyche, a fact which reveals that she thinks of the work involved in managing my mental life as highly complex and energy consuming when left to myself. She envisages the possibility of total depletion and has the experience of being unable to move forward for lack of energy and/or light. In this space, grace enlivens, because it is an encounter with a "pure" person, who draws me into Their personal sphere, and—to the extent that I am open for it though faith—fills me with Their Spirit, to guide my motivational life. This Spirit "fulfils" the person with supernatural life power (life, simply), as it moves towards an adequate value response to everything, it being divine. Agency in relation to this form of social unity with God is retained by the person in the prerogative of faith (Lebech, 2023b). While union with God thus relativises all other types of social dependencies by means of faith, it does not suspend the phenomena. On the contrary, the phenomena help Stein to phenomenologically describe the experience of grace as the subsumption of the person into divine community.

9.5 Conclusion

Psychic causality in its relation to motivation—as conceptualised by Stein—allows us to analyse the structure of social experience in its diverse dimensions. It can—with its corollary, social causality—serve the social sciences to identify their object of study in its dependency on personal motivation and intersubjective constitution. It would leave room for those sciences to capture the role played by values in social life, in so far as social causality both conditions value response and also essentially involves it, given that no act of an individual person (whose exposure to psychic causality also exposes him to social causality) can be accomplished without it being motivated. Social science, concerned with identifying what it is that conditions people in their value response, is not interested in the value response itself, but in how it exemplifies or is conditioned by social trends, tendencies, developments, and movements (possibly in order to effectively realise goals by means of deploying social causality). It makes sense to thematise this conditionality as social experience's dependency on social causality. It should therefore be further explored whether this thematization can advance the goals of the social sciences. Management studies might lead the way, since that field has recognised how personal and communal engagement with values may function as means to effectively realise goals.

References

Beckmann-Zöller, B., & Gerl-Falkovitz, H.-B. (Eds.). (2006). *Die unbekannte Edith Stein: Phänomenologie und Sozialphilosophie*. Peter Lang.

Betschart, C. (2011). *Was ist Lebenskraft? Eine Auseinandersetzung mit. Edith Steins Untersuchung „Psychische Kausalität"* (Licenciate thesis). Pontifical University Gregoriana, submitted 2008, revised 2011.

Burns, T., & Vendrell Ferran, I. (2024). *Edith Stein's on the problem of empathy: A companion*. Bloomsbury.

Calcagno, A. (2014). *Lived experience from the inside out. Social and political philosophy in Edith Stein*. Duquesne University Press.

Feldes, J. (2015). *Das Phänomenologenheim. Der Berzabener Kreis im Kontext der frühen phänomenologischen Bewegung*. Verlag Traugott Bautz.

Gaudiano, V. (Ed.). (forthcoming). *Edith Stein's contributions to the philosophical foundation of psychology and the humanities: A companion*. Bloomsbury.

Hagengruber, R. (2006). Sozialphilosophie als "strenge Wissenschaft" – Überlegungen zu Edith Stein und Edmund Husserl. In C. Beckmann-Zöller & G. Gerl-Falkovitz (Eds.), *Die unbekannte Edith Stein: Phänomenologie und Sozialphilosophie*. Peter Lang.

Hall, J. (1977). Alfred Schutz, his critics, and applied phenomenology. *Philosophy and Social Criticism, 4*(3), 265–279. https://doi.org/10.1177/019145377700400303

Husserl, E. (1960). *Cartesian meditations* (D. Cairns, Trans.). Martinus Nijhoff.

Husserl, E. (1991). *On the phenomenology of the consciousness of internal time (1893–1917)* (J. B. Brough, Trans.). Kluwer.

Husserl, E. (1992). Cartesianische meditationen. In *Krisis, Gesammelte Schriften* (Vol. 8). Felix Meiner.

Husserl, E. (2003). *Phänomenologische psychologie: Vorlesungen sommersemester 1925*. Felix Meiner.

Lebech, M. (2004). Study-guide to Edith Stein's philosophy of psychology and the humanities. In M. Lebech (Ed.), *Yearbook of the Irish philosophical society* (pp. 40–76). Maynooth. A substantially revised version is found in Gaudiano, forthcoming.

Lebech, M. (2009). *On the problem of human dignity: A hermeneutical and phenomenological investigation*. Königshausen und Neumann.

Lebech, M. (2015). Motivation and value. In *The philosophy of Edith Stein: From phenomenology to metaphysics* (pp. 27–40). Peter Lang.

Lebech, M. (2017). Stein's understanding of mental health and mental illness. In E. Magri & D. Moran (Eds.), *Empathy, sociality, and personhood: Essays on Edith Stein's phenomenological investigations* (pp. 107–123). Springer.

Lebech, M. (2019). Essence, eidos, and dialogue in Stein's "Husserl and Aquinas: A comparison". In T. Burns, T. Szanto, & A. Salice (Eds.), *The new yearbook for phenomenology and phenomenological philosophy* (Vol. 17, pp. 22–36). Routledge.

Lebech, M. (2023a). Edith Stein (1891–1942). In N. de Warren & T. Toadvine (Eds.), *Encyclopaedia of phenomenology*. Springer. https://link.springer.com/referenceworkentry/10.1007/978-3-030-47253-5_76-1

Lebech, M. (2023b). Stein's phenomenology of grace. *Religions, 14*(7), 950. https://doi.org/10.3390/rel14070950

Lebech, M. (forthcoming). Stein's reading of Scheler's *Idole der Selbsterkenntnis* (1915): The centrality of the phenomenological method and the nature of the psyche. In T. A. Burns, T. Lacy, & E. Mohr (Eds.), *Stein and Scheler in dialogue*. Bloomsbury.

McGuirk, J. (forthcoming). Influence. In V. Gaudiano (Ed.) *Edith Stein's* Philosophy of Psychology and the Humanities.*A Companion*. Bloomsbury.

Moran, D. (2004). The problem of empathy: Lipps, Scheler, Husserl, and Stein. In T. A. F. Kelly & P. Rosemann (Eds.), *Amor amicitia: On the love that is friendship. Essays in medieval thought and beyond in honor of the Rev. Professor James McEvoy* (pp. 269–312). Peeters.

Overgaard, S. (2019). What is empathy? In F. Kjosavik, C. Beyer, & C. Fricke (Eds.), *Husserl's phenomenology of intersubjectivity: Historical interpretations and contemporary applications* (pp. 178–192). Routledge.

Peritore, N. P. (1975). Some problems in Alfred Schutz's phenomenological methodology. *American Political Science Review, 69*(1), 132–140. https://doi.org/10.2307/1957890

Rang, B. (1973). *Kausalität und Motivation: Untersuchungen zum Verhältnis von Perspektivität und Objektivität in der Phänomenologie Edmund Husserls*. Martinus Nijhoff.

Riess, W. (2010). *Der Weg vom Ich zum Anderen. Die philosophische Begründung einer Theorie von Individuum, Gemeinschaft und Staat bei Edith Stein*. Thelem.

Sawicki, M. (2003). The humane community: Husserl versus Stein. In R. Feist & W. Sweet (Eds.), *Husserl and Stein* (pp. 141–154). The Council for Research in Values and Philosophy.

Schutz, A. (1962). Husserl's importance for the social sciences. In *Collected papers I* (pp. 140–149). Martinus Nijhoff.

Schutz, A. (1967). *The phenomenology of the social world* (G. Walsh & F. Lehnert, Trans.). Northwestern University Press.

Schutz, A. (1975a). Edmund Husserl's ideas. In *Collected papers III* (Vol. II, pp. 15–39). Martinus Nijhoff.

Schutz, A. (1975b). Type and *eidos* in Husserl's late philosophy. In *Collected papers III* (pp. 92–115). Martinus Nijhoff.

Stein, E. (1989). *On the problem of empathy. CWES III* (W. Stein, Trans.). ICS Publications.

Stein, E. (2000). *Philosophy of psychology and the humanities. CWES VII* (M. C. Baseheart & M. Sawicki, Trans.). ICS Publications.

Stein, E. (2006). *An investigation concerning the state. CWES X* (M. Sawicki, Trans.). ICS Publications.

Stein, E. (2014a). *Freiheit und Gnade und weitere Texte zu Phänomenologie und Ontologie (1917–1937). ESGA IX*. Herder.

Stein, E. (2014b). *Letters to Roman Ingarden. CWES XII* (H. Hunt, Trans.). ICS Publications.

Szanto, T., & Moran, D. (2025). Edith Stein. In E. N. Zalta & U. Nodelman (Eds.), *Stanford encyclopedia of philosophy*.

Mette Lebech is Associate Professor of Philosophy at Maynooth University, Ireland, since 1998. She holds degrees from the universities of Copenhagen, Louvain-la-Neuve and Leuven. She has published widely on human dignity and the philosophy of Edith Stein. Her publications include *On the Problem of Human Dignity. A Hermeneutical and Phenomenological Investigation* (Königshausen und Neumann, 2009); *The Philosophy of Edith Stein. From Phenomenology to Metaphysics* (Peter Lang, 2015); *European Sources of Human Dignity. A Commented Anthology of Texts on Human Dignity* (Peter Lang, 2019) and *Value Phenomenology. Taking Account of Edith Stein's Contributions* (Bloomsbury, 2026). She is founding President of the International Association for the Study of the Philosophy of Edith Stein (IASPES), Series Editor of *Edith Stein Studies*, Bloomsbury and a Fellow at the Centre for the History of Women Philosophers and Scientists at the University of Paderborn.

Part IV
Transcendental Intersubjectivity and Philosophical Anthropology

Chapter 10
Stein's Philosophical Anthropology as a Completion of Husserl's Project for a Foundation of the Sciences. *Ideen II* – The "Master" and His Co-Genius

Claudia Mariéle Wulf

Abstract Edmund Husserl and Edith Stein seem to share the same starting point: the pure I. Though Stein's epistemology is based on this concept, her eidetic phenomenology will include the quality of the personal I. This fundamental decision has a striking influence on her vision of the object: it is more than its mere representation in the consciousness, it has an ontic quality and touches the ontic me, the concrete person. The quality of the subject is crucial for how it is entangled in the concrete reality. Stein explores this, for example, in her work "Psychic Causality". Existential questions gradually become more visible in her research. Nevertheless, she sticks to Husserl's epistemology: the concept of the subject must be as clear as the object, and the notions must be clarified in order to understand the essence of what will be subjected to further research. Eidetic phenomenology will, according to Edith Stein, be the foundation of all further empirical and qualitative research.

Keywords Anthropology · Eidetic phenomenology · Epistemology · Existentials · The sciences of the humanities · Stein · Husserl

Edith Stein: I regard her as Edmund Husserl's co-genius. He was a master, no doubt. Thus, his (in the end, transcendental) phenomenology[1] was further elaborated by Edith Stein's vision, the so-called "eidetic phenomenology".[2] Husserl's aim was to constitute a precise phenomenology[3] which follows the ideal of mathematics. The threefold *epoché* and the phenomenological reduction call for a radical change of

[1] Husserl (2025, p. 611). Husserl regards "transcendental subjectivity" as the origin of meaning.

[2] Husserl also claims this notion for his view, though he clearly argues transcendentally (Husserl, 2025, p. 614).

[3] Husserl (2025, p. 610f.): Husserl meant it to be a "strict science".

C. M. Wulf (✉)
Independent Scholar, St. Gallen, Switzerland
e-mail: mariele.wulf@bluewin.ch

M. Lebech (ed.), *Stein and Husserl's Phenomenology of the Social in Ideas II*, Women in the History of Philosophy and Sciences 25,
https://doi.org/10.1007/978-3-032-17339-3_10

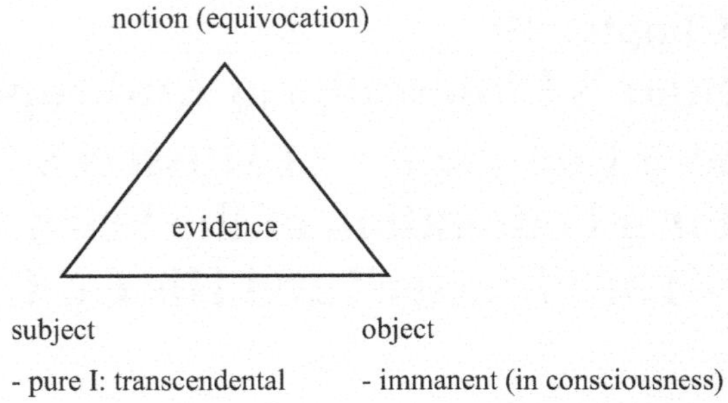

Fig. 10.1 Husserl's vision of evidence

attitude.[4] The "transcendental apriori" (Husserl, Ideen II, p. 615) is the *conditio sine qua non* for his new view. Though this way of thinking puts existence into brackets, it even touches on metaphysics.[5] Husserl strived for the comprehensive constitution of the object in the subject; Stein investigates the constitution of the human person, the analysing subject *as* a subject of its own. Husserl fears a return to the psychologism he tried to overcome, and therefore he chooses the way of transcendental phenomenology.[6] According to Stein, like others of her generation, this step is not necessary.

Husserl strives for "*evidence*": the absolute clear view of the essence, available by intuition (cf. Husserl, Ideen II, p. 435). His vision can be illustrated by the triangle of evidence (Fig. 10.1).

Husserl searches for clear notions, beyond what he called "equivocation": similar words used for expressing different concepts. Unclear notions lead to unclear research. In order to find the right notion, the subject and the object must be as ideal as possible. Husserl chooses the "pure I", the subject beyond psychologism, as subject of knowledge and the immanent object, the object "represented"—i.e., given in the consciousness—as the ideal object of sciences.[7]

10.1 Transcendental Versus Eidetic Phenomenology

Like Husserl, Edith Stein strives for the precise notion, which expresses the essence. But she considers another kind of subject and object.

[4] Husserl (2025, pp. 616–7): "*Einstellungsänderung*" from the natural to a phenomenological attitude. The „Zugangsmethode zur transzendental-phänomenologischen Sphäre"is „phänomenologische Reduktion"(Husserl, 2025, p. 613).

[5] Husserl (2025, p. 616f.): Existence does not play a role anymore. Cf. Husserl (2025, p. 612f.): Transcendental phenomenology touches the universal and metaphysical horizon.

[6] Cf. Husserl (2025, p. 618, p. 620f.): He fears a „Rückkehr zur natürlich psychologischen Einstellung".

[7] „Bewusstsein" and „Bewusstseinsrepräsentation"in the German original.

10.1.1 The Points of Departure: The Pure I and the Personal I

10.1.1.1 The Pure I as Point of Departure for an Epistemological Approach

In Husserl's construction, the *pure I* is the point of departure and the centre of the stream of conscience (cf. Husserl, 2025, p. 5f. and p. 318). In phenomenological language, Husserl assesses the real I, the complex person, as "transcendent". The transcendental, or pure I, is immanent (cf. Husserl, 2025, p. 10). This pure I is the *centre* of conscience and the pole of actuality (cf. Husserl, 2025, p. 305ff.). It may take the not-yet-realized thoughts into account, but in its thoughts, the I becomes aware of itself (Husserl, 2025, p. 7 and p. 310). The I constitutes itself via the empirical I and is able to reflect itself and its cogitations (Husserl, 2025, p. 178f., p. 205, p. 309 and p. 315). It is identical, individual and free of all qualities (Husserl, 2025, p. 231, p. 312, p. 349ff.) and constitutes the identity of the stream of conscience (Husserl, 2025, p. 27, p. 339ff.). The pure I is also the freely acting subject—like the subject of the Cartesian cogito (Wulf, 2005, p. 147ff.; Husserl, 2025, p. 306f.; Husserl, 1950, p. 75). This acting has a specific personal quality, which includes feelings, freedom, and even bodily experiences (Husserl, 2025, p. 205, p. 248 and p. 307f.). Though it reflects the concrete surroundings and all its qualities, this I remains *transcendental*.[8]

The *real I*, however, is a *transcendent* reality (Husserl, 2025, p. 8ff. and p. 317ff.). Husserl first thinks that the pure I has access to the personal I (Husserl, 2025, p. 27ff.). Later, he corrects himself and states: the identity is rooted in the pure I, the real I is the counterpart of the pure I (Husserl, 2025, p. 5 and p. 30). The pure I constitutes the real I; the latter seems to be part of its "surrounding", its experiences and even its objects (Husserl, 1952, p. 114; 2025, p. 7f., p. 34, p. 224 [110], p. 315ff., p. 544 and p. 550ff; 1989, I, p. 363.). The real I is bound to the body; it constitutes a kind of "materiality" of the soul (Husserl, 2025, p. 10f., p. 15f., p. 228 and p. 298ff.). The soul is a complex entity with different layers, given to the pure I as a transcendent reality like the body; its unity is constituted by the experiencing I (Husserl, 2025, p. 13, p. 16f., p. 18f. and p. 22). The real I is a psychological entity, the subject of the stream of experience (Husserl, 2025, p. 34 and p. 220ff.). The real I, its body, and all surroundings are constituted by the pure I, but in the immanent sphere, i.e., that of consciousness (Husserl, 2025, p. 42, p. 55f. and p. 317ff.). The question of how this constitution really takes place remains an open one (Husserl, 2025, p. 552). Stein calls this the "irrational remain" (Stein, 2005, p. 236)[9] of transcendental phenomenology

---

[8] Husserl (2025, p. 27 and p. 305). It even constitutes a transcendental intersubjectivity: Cf. Husserl (2025, p. 625).

[9] See also Stein (2004a, p. 166): „Ein dem kantischen verwandter transzendentaler Idealismus, der den Aufbau der äußeren Welt als Formung des sinnlichen Materials durch Kategorien versteht, erscheint bestenfalls als eine mögliche Deutung des phänomenalen Bestandes, ohne einen Beweis der Daseinsrelativität der Außenwelt geben zu können. Auf die Frage nach der Herkunft des ichzugehörigen und doch ichfremden Materials muss er die Antwort schuldig bleiben. Er behält einen

as it is not clarified. The personal I, which constitutes itself as a person, cannot be known with the same undeniable clarity as the pure I.[10]

10.1.1.2 The Personal I as Point of Departure in Eidetic Phenomenology

Methodologically, Stein chooses the same point of departure (the immanent pure I), which has no quality and is not part of the (transcendent) world.[11] It is the unifying point of conscience, an "absolute being" (Stein, 2004b, p. 73). But Stein regains the personal I as a starting point (Wulf, 2005, Chapter 4); she sees that we meet ourselves as psychophysical individuals in consciousness itself (Stein, 2004b, p. 88). Thus, the personal quality of the I is given in transcendental reduction. The gap between the transcendental consciousness (immanence) and the transcendent world can be overcome in the personal consciousness, where the I comes to understanding of itself (Stein, 2004b, p. 105; 2001, p. 163 [90/28.9.25]).

In Stein's vision, the *acting* I is the *personal* I: the I between immanence and transcendence Stein, 2004b, p. 67; 2006, p. 56. The pure I without any quality cannot experience the quality of the transcendent world, as Stein tells us (Stein, 2008, p. 106). The original consciousness of the personal I is more than the stream of consciousness (Stein, 2004b, p. 176ff.; 2005, p. 80 and p. 83 [85]). Its memory is more than mere "retention"; it enters the deeper layers of the soul (Stein, 2006, p. 368f. [417/418]). The stream of consciousness is related to the personal structure of the I, to the soul where the personal I lives (Stein, 2008, p. 56; 2004e, p. 86; 2006, p. 362ff.). This I comprehends the quality of the soul and it is the real centre of acting (Stein, 2010, p. 258). In this regard, Stein differs from Husserl.[12]

---

irrationalen Restbestand. Und schließlich wird man sagen müssen, dass er den Phänomenen nicht gerecht wird: die Wesens- und Daseinsfülle, die in aller echten Erfahrung in das erfahrende Subjekt einbricht und bewusstseinsmäßig alle Fassungsmöglichkeiten übersteigt, widerspricht der Rückführung auf eine blosse Sinngebung vom Subjekt her. So scheint mir gerade die getreue Analyse der Realitätsgegebenheit zu einer Aufhebung der transzendentalen Reduktion und zu einer Rückkehr in die Haltung der gläubigen Hinnahme der Welt zu führen."

[10] Husserl (2025, p. 35ff. and p. 551ff). Husserl sees a certain parallelism in the relation body-soul and soul-spirit.

[11] Stein (2008, p. 11; 2004b, p. 20): The pure I „ist nicht ein Stück der realen Welt wie das psychische Individuum, sondern steht der Welt gegenüber".

[12] Husserl (2025, p. 319, p. 516). Husserl seems to attribute this kind of acting to the pure I.

10.1.2 The Different Vision of the Object

10.1.2.1 Husserl's Vision: The Immanent Object of the Pure I

Husserl describes two kinds of perception: the object is given (1) in phenomenological reduction or (2) by concrete experience (Husserl, 2025, p. 34 and p. 544f.). Husserl splits the immanent sphere, the sphere of the essence (*a priori*), from the transcendent sphere, the concrete object (*a posteriori*) (Husserl, 2025, p. 139f.). Husserl transfers each object into the immanent sphere (constitution) (Husserl, 2025, p. 25), as the ideal object is the immanent object, which cannot be questioned (Husserl, 2025, p. 8). The *noema* is the thing to be analysed, rather than the essence itself (Husserl, 2025, p. 150); the essence appears in *noesis*. *Noesis* can be regarded as the process of the constitution and understanding of the object (values, persons and all experiences) in the immanence of the consciousness (Husserl, 2025, p. 54f., p. 85, p. 204 and p. 549). The analysis of the person is seen in this context: the constitution of the person is not a "miracle", but rather a rational solution for the epistemological problem given in the process of constitution (Husserl, 2025, p. 139ff.).

10.1.2.2 Stein's Vision of the Object

The quality of the I meets with the quality of the object. Stein uses Husserl's methodology, his threefold *epoché*, and analyses the representation of the object in consciousness (Stein, 2004d, p. 149; 2004b, p. 15, p. 19 and p. 25f.). This is the first step of the knowledge of essence. But she does not forget about the concrete object, the existing thing.[13] Its "ontic structure" (Stein, 2010, p. 163ff.; 2004b, p. 113ff.; 2006, p. 210) is never ideal, but during the constitution it is perceived, intentionally given, and will thus be interpreted (Stein, 2004b, p. 18, p. 66ff. and p. 86f.; 2010, p. 40f.). In addition, the object has a relevance for the subject, a value. Thus, the second step is the constitution of value (Stein, 2010, p. 134). The third step is intersubjective constitution, also referred to by Husserl: In intersubjective comparison, the essence is confirmed (Stein, 2010, p. 131. Husserl, Ideen II, p. 190; 1975, p. 262, editor's introduction).

The constitution of the values is of special interest in this context. Husserl treats axiology in a formal way (Stein, 2006, p. 273). Stein, however, sees the value in the thing as such, the good (Stein, 2006, p. 272f.; 2004e, p. 82; 2004b, p. 129; 2010, p. 136). Values are given to the person, and they ask for a personal answer (Stein, 2008, p. 25 [15]; 2010, p. 190). Only a subject with personal qualities is able to "feel" values; a capacity that the pure I does not have (Stein, 2006, p. 421; 2010, p. 64f., p. 134ff.; 2008, p. 119).

---

[13] Stein (2004d, p. 150; 2004b, p. 90; 2006, p. 149). Sepp (n.d., p. 19).

10.1.3 *The Clear Notion of the Essence*

The aim of phenomenology is to reach the clear concept beyond all equivocation. Here, transcendental and eidetic phenomenology meet! Husserl wants to clarify the notions which express the essence (Husserl, 2025, p. 161ff. and p. 164). The aim is the "completing assumption"[14]: the clear notion that covers the full knowledge of the essence. This notion must even cover all deep dimensions of the concept—a quality which is, according to Stein, only given in the personal I and not in the pure I. The concrete application of this will be treated later (10.3.1.2).

10.2 Epistemological and Existential Questions

10.2.1 *Describing the Essence: Positive Phenomenology*

10.2.1.1 Husserl's Positive Phenomenology: Finding the Essence

The phenomenological reduction goes in both directions: it reduces the object to its representation in consciousness, and it reduces the I to the pure I; with the aim, in both cases, of finding the true concept (Husserl, 2025, p. 141f). The transcendental consciousness is the starting point for all further knowledge, even for the exact sciences and psychology (Husserl, 2025, p. 143f.). All "transcendent" being is put into brackets. What remains is consciousness as such. This pure consciousness is, according to Husserl, the only place for knowing the essence of the object and for knowing yourself (Husserl, 2025, p. 141 and p. 147ff.). This could be interesting for psychology.

10.2.1.2 Stein's Positive Phenomenology

Stein applies Husserl's methodology, but she does not follow his transition to transcendental perception (Stein, 2004d, p. 152). She sticks to ontology; to the realistic view. Empirical psychology, however, chooses the other extreme: here, the essence plays no role anymore and basic concepts are absent. This is why phenomenology became attractive to Edith Stein, as it analyses the notions and, by doing so, creates a valid basis for empirical research (Stein, 2002, p. 174). By clarifying the basic notions of anthropology, Stein laid the foundations for education as a science (Stein, 2004e, p. 2f.).

---

[14] Husserl (2025, 165ff.): „Erfüllende Anschauung“

10.2.2 The Absence of the Essence: Negative Phenomenology on "Psychic Causality" and Freedom

Stein and Husserl mainly strive towards clarifying the essence by applying phenomenological reduction, the threefold *epoché*, and by exercising phenomenological variation. An important step during this clarification is the negation of certain essential traits in order to see whether they are part of the essence or not. I will apply this to freedom as an essential trait of the human being. Both Husserl and Stein deal with this question.

10.2.2.1 Husserl's Description of "Psychic Causality" and Freedom

The pure I is free; it can be active or non-active, but by mere free will, without being motivated from the outside. But concrete persons live in concrete circumstances, in causalities that they cannot leave and which shape them (Husserl, 2025, p. 15, p. 323). Husserl describes psychology as exact science and even constructs a parallelism between the material reality and the human soul (Husserl, 2025, p. 14, p. 188f., p. 469). The motivation in the soul itself, however, the motivation of the person, is not caused by any circumstances; it is rational by nature (Husserl, 2025, p. 453f., p. 474f.).

To understand persons means to intuitively follow their life and to understand the "nature" of their motivation (Husserl, 2025, p. 458, p. 468). Values can be motivating if a person decides to accept and implement them (Husserl, 2025, p. 519, p. 531f.). Here, the subject reappears as a free person—beyond strict causality (Husserl, 2025, p. 520, p. 530f., p. 540). For this reason, Husserl finally discerns psychophysical causality from physical causality (Husserl, 2025, p. 539). The I is the subject of its acts; it is characterized by rationality and autonomy (Husserl, 2025, p. 584ff). Therefore, the person, whose will Husserl identifies with the acting pure I, is responsible (Husserl, 2025, p. 516ff. and p. 351f.). Psychology must understand that all experiences are the experiences of this free and complex person (Husserl, 2025, p. 517, p. 585, p. 630). For this reason, exact sciences must be incorporated into the human sciences, and their attitude must be transcendental (Husserl, 2025, p. 470 and footnote 2 [remark by the editor]).

10.2.2.2 Stein's Analyses of "Psychic Causality" and Freedom

Causality is the absence of freedom. In motivation, freedom as essential trait of the human being must be present. According to Edith Stein, an empirical psychology overlooks the double law of causality and motivation (Stein, 2010, p. 5ff., p. 68; 2004b, p. 118). Stein clarifies this in her research on "psychic causality", which is a first step to a humanistic psychology (Stein, 2010, p. 5f. [3]). The free subject initiates the act; said subject can assume or suppress the influences as the free will can break the law of causality (Stein, 2004e, p. 127; 2004b, p. 121f.; 2010, p. 84, p. 54ff., p. 78f.). Sometimes it seems as if the psyche is dominated by causality which cannot be influenced (Stein, 2010, p. 16ff., p. 29f.). Stein does research on this presumed causality: it does lie in the laws which reign within the person, but the *a priori* freedom of the will can overcome these (Stein, 2010, p. 26; 2004c, p. 16; 2004e, p. 6). Thus, the reaction of the person cannot be calculated (Stein, 2010, p. 30). The free will characterises the "causality" in the psyche as motivation (Stein, 2004b, p. 121; 2004e, p. 38f.; 2006, p. 370; 2010, p. 6, p. 41): the spirit initiates a new causality by possible meaning; motivation stimulates an act but does not force the person to react this way (Stein, 2004b, 2004e, p. 91; p. 210f.; 2006, p. 336; 2010, p. 98). In this, Stein is clearer than Husserl and Scheler. This negative phenomenology, which eliminates the possible freedom, shows one thing: freedom is—undeniably—an essential trait of the human person. The implication for a possible psychology is immediately clear: freedom from psychic causality, due to a psychical wound or a difficult past, must be an essential aim of therapy.

10.2.3 Negative Phenomenology in Experience: Existential Dimensions e.g. Freedom

10.2.3.1 A priori Freedom Versus Existential Freedom

Stein now goes beyond the sober analysis of a negative phenomenology and touches on the existential dimension, the dimension given in the human experience. At this point, her decision to keep the personal I as the acting subject proves fruitful. This personal I has access to its feelings, to the values and to the person's dignity. Freedom or the lack of freedom are existential experiences, as she shows in the text "Freedom and Grace" by analysing the possibility of absolute freedom. The I, leaving behind all restrictions, gains total freedom: the transcendental freedom. However, the I which gave up all concrete possibilities faces total emptiness; in this freedom, the I would be totally bound and totally fixed to the point of absolute freedom (Stein, 2004c, p. 13). These transcendental analyses end up in an existential paradox. The existential experience, however, is different: the point of absolute freedom is the starting point when we realize our concrete freedom. Freedom exists; it is only not used if you do not make any choice (Stein, 2004c, p. 13). The aim of the free act, the concrete realization of meaning, does not exist in the transcendental, immanent sphere, but rather only in the ontological, transcendent sphere. Freedom is not experienced as a transcendental trait, but in a concrete free action, always initiated by our will, whether it's a free thought, a free word or a free action (Stein, 2005, p. 108).

This existential analysis of freedom is thus deeper than the rational analyses of transcendental philosophy. It recurs to the experience of freedom in a positive way and, by this, touches the existential sphere where meaning rises.

10.2.3.2 Emptiness and Fullness

The relationship to a presumed meaning in life correlates with two *existentials*: with fullness in the case of a meaningful relationship, and with emptiness in the case of a destructive relationship. If a person wants to be free from all influences, all contents, in order to become totally free, then nothing remains but emptiness (Stein, 2004c, p. 17; 2006, p. 52, p. 56, p. 318; 2010, p. 65). Fullness can only be found beyond the pure I: in the soul where it is at home and in its experiences where it realises its mere being and where it finds meaning and fullness (Stein, 2004c, p. 17; 2006, p. 57f.). At that very moment, the human being touches another fullness: the One who carries his being (Stein, 2006, p. 58). Human beings are afraid of not being. However, they do not strive to live forever, but rather to live the fullness of life, ultimately provided by God (Stein, 2006, p. 58, p. 385).

10.3 Eidetic Phenomenology as a Foundation of Humanistic Sciences

Phenomenology can provide a threefold service to exact sciences: it can clarify the notions; it can stimulate thought beyond existing concepts; and an existential phenomenology can testify to the undeniable essence of the human being. This will shortly be indicated for psychology.

10.3.1 Clarifying the Notions

10.3.1.1 Husserl on the Foundation of the Sciences

The first aim of a phenomenological foundation of a science is *the clarification of the essence of the science itself and of its object*. Its key notions cannot be analysed by the scientists themselves; this task remains in the hands of the philosophers, especially of phenomenologists (Husserl, 2025, p. 76. Footnote 1). Phenomenology is the last resort; beyond it, no place can be found for this kind of analysis. It provides insight into the essence of the science and its object and into the essence of its methodology (Husserl, 2025, 157, p. 418). Pure phenomenology and humanistic psychology do research on this subject (Husserl, 2025, p. 476). Eidetic phenomenology deals with concrete reality—as opposed to a phenomenology applying the transcendental reduction (Husserl, 2025, p. 502).

The next step will be to *clarify* the *methodology*. The phenomenological description prepares empirical explanations; both aim at objectivity (Husserl, 2025, p. 489). Phenomenological analyses and the description of the essence should precede each application (Husserl, 2025, p. 125, p. 134). Phenomenology also discovers essential possibilities within a particular science; this is why these analyses can provide an objective foundation for humanistic sciences (Husserl, 2025, p. 121, p. 437).

If we apply this to psychology as a humanistic science, each research effort in this field should be preceded by clear analyses of the person: It is *an I, a subject*, having a spirit, living in relationship with the surroundings and with other persons. All these relationships should be examined and fully realised (Husserl, 2025, p. 490f.). The fact that the person has a self-apperception, a self-value and a self-willingness, that he/she can be experienced and understood, is crucial for developing a well-founded psychology (Husserl, 2025, p. 501f.). Researchers in psychology must understand the essence of the subject's relatedness to others and to otherness in freedom and spirit; the way in which the free I appears and acts in the person, and how it constitutes the unity of all experiences (Husserl, 2025, p. 492, p. 82, p. 105, footnote). This helps to understand the complex process of motivation and empathy (Husserl, 2025, p. 489ff.).

10.3.1.2 Stein's Vision of Founding Sciences

Stein also provides a vision on the analyses of sciences. She strives for a foundation of a humanistic psychology as a "descriptive or describing psychology" (Stein, 2004b, p. 19; 2010, p. 3) which differs from explaining psychology. It seems to be forbidden to speak about the sounder reality, the "soul", in psychology (Stein, 2006, p. 522). The task of phenomenology is to clarify this notion next to the basic concepts of an "aprioric" psychology (Stein, 2004b, p. 21 [36]; 2010, p. 254). This is precisely what Stein strives for in her *Beiträge zur Philosophischen Begründung der Psychologie und der Geisteswissenschaften.* A psychology without soul, skipping the very concept of "soul", does not know about its own essence and needs to be underpinned by phenomenological analyses of the essence of the soul (Stein, 2006, p. 23, p. 521; 2010, p. 101f., [108]. p. 254, p. 246ff.). Otherwise, psychology risks restricting itself to empirical research, which merely describes causalities and the "mechanism of the soul" (Stein, 2010, p. 103) but loses sight of the living reality (Stein, 2006, p. 521; 2010, p. 101).

10.3.2 Beyond Existing Concepts: Thinking Outside the Box

In psychology, thinking in "schools" is very common. One of the main diagnostic instruments, the DSM, is based on fixed concepts on what a psychical disease is. Several symptoms are enumerated and translated into questions or statements that the patient will find in a questionnaire. The main problem is that these concepts are not questioned anymore! Once fixed, the concept of "depression", "personality disorder" etc., seems to be determined forever. However, the same symptoms can appear in different contexts. Being tired and without energy might be a sign of depression, but it is not its essence! Panic attacks after severe events and breaks in life are immediately linked to trauma. With research on trauma, based on Edith Stein's model of the soul, one can show that the essence of trauma cannot be brought back to severe incidents alone, but that micro-traumatisation has the same effect: namely, the emergency shutdown of the soul. This is the main essential trait. By phenomenological analyses, it is possible to show that trauma often appears in close relationships—even if abuse or violence cannot be seen on the outside. In relationships on which one depends, one can easily be destroyed—such as by rape or murder (Wulf, 2021). It would be a revolution for psychology, as well as in juridical contexts, to apply this concept of trauma. Stein provided the groundwork for this thinking.

Psychology should be based on a comprehensive anthropology. The concepts of personhood differ in the different schools, as they are not able to provide philosophical and metaphysical analyses. In this regard, phenomenology can play a decisive role!

10.3.3 The "Existentials" as Source of Undeniable Knowledge

The problem we must deal with in the context of a psychology that acts according to the standards of exact sciences is that we have to prove the humanistic insights in a convincing way. We saw that Stein has an existential approach in phenomenology. *Existentials* are essential relationships, which are revealed by existential experiences.

- Our identity is essentially linked to our I; during trauma, this relationship breaks.
- Individuality is our personal complexity, which may be broken by psychological rape.
- Our intellect needs truth; otherwise, we lose our capacity to think logically.
- Our (living) body is bound to material resources (place and time); if we don't get that (like the aborted unborn who didn't get its time), we die!
- Our feelings, deeply rooted in our soul, need orientation in values and the experience of dignity; otherwise, we lose ourselves and our humanity.
- We are social beings and need relationships; without them, we will shrink.
- By language, willingly possible expressions such as words, gestures, behaviours and mimic, we can show that we are human, that we can think and love and that we can claim our rights; without which we would be lost.
- Morality needs to be orientated to freedom and values; otherwise, we lose ourselves in inhuman behaviour and ultimately hurt others.
- Transcendence must be orientated to a comprehensive meaning—ultimately, to a transcendent person. Meaninglessness destroys our life.

Human beings cannot live without these relationships, as their essential traits become visible and develop in these relationships (Stein, 2004b, p. 170; 2006, p. 425; 2010, p. 222f. Wulf, 2010). Existential relationships are thus a proof for human essence.

10.4 A Critical Question: Edith Stein in "Ideen II"

I will conclude with a critical question: How much of Edith Stein can be found in the "*Ideen II*"? Some passages have the same diction as in her books. However, the point of departure remains different. Husserl declares it clearly in the end of this book. He sticks to the pure I, to the transcendental point of view, and does not want to be misunderstood as someone who may be suspected of psychologism. This is his main concern. But in this way, he misses the chance to make his phenomenology fruitful for the human sciences, including psychology. The existential experience is excluded. This is what Edith Stein definitely added, and this is her undeniable merit.

References

Husserl, E. (1950). *Ideen zu einer reinen Phänomenologie und phänomenologischen Philosophie I, Hua III*. Martinus Nijhoff.

Husserl, E. (1952). *Ideen zu einer reinen Phänomenologie und phänomenologischen Philosophie III, Hua V*. Martinus Nijhoff.

Husserl, E. (1975). *Logische Untersuchungen. Erster Band. Prolegomena zur reinen Logik, Hua XVIII*. Martinus Nijhoff.

Husserl, E. (1984). *Logische Untersuchungen. Zweiter Band. Erster Teil, Hua XIX/1*. Martinus Nijhoff.

Husserl, E. (2025). *Ideen zu einer reinen Phänomenologie und phänomenologischen Philosophie. Zweites Buch: phänomenologische Untersuchungen zur Konstitution und Wissenschaftstheorie. Die drei Teilentwürfe und Thematische Ausarbeitungen sowie das „Nachwort zu meinen Ideen" (1908–1930)* (D. Fonfara, Ed.). Springer.

Sepp, H. R. (n.d.). *Edith Steins Stellung innerhalb der Phänomenologischen Bewegung*. Unveröffentlichte Maschinenschrift.

Stein, E. (2001). *Selbstbildnis in Briefen III. ESGA 4*. Herder.

Stein, E. (2002). *Aus dem Leben einer jüdischen Familie. ESGA 1*. Herder.

Stein, E. (2004a). *Diskussionsbeiträge Juvisy. In "Freiheit und Gnade" und weitere Beiträge zu Phänomenologie und Ontologie (1917 bis 1937). ESGA 9* (pp. 162–167). Herder.

Stein, E. (2004b). *Einführung in die Philosophie. ESGA 8*. Herder.

Stein, E. (2004c). Freiheit und Gnade. In *"Freiheit und Gnade" und weitere Beiträge zu Phänomenologie und Ontologie (1917 bis 1937). ESGA 9* (pp. 8–72). Herder.

Stein, E. (2004d). Die weltanschauliche Bedeutung der Phänomenologie. In *"Freiheit und Gnade" und weitere Beiträge zu Phänomenologie und Ontologie (1917 bis 1937). ESGA 9* (pp. 143–158). Herder.

Stein, E. (2004e). *Der Aufbau der menschlichen Person. Vorlesung zur philosophischen Anthropologie. ESGA 14*. Herder.

Stein, E. (2005). *Potenz und Akt. Studien zu einer Philosophie des Seins. ESGA 10*. Herder.

Stein, E. (2006). *Endliches und ewiges Sein. Versuch eines Aufstiegs zum Sinn des Seins. ESGA 11/12*. Herder.

Stein, E. (2008). *Zum Problem der Einfühlung. ESGA 5*. Herder.

Stein, E. (2010). *Beiträge zur philosophischen Begründung der Psychologie und der Geisteswissenschaften. ESGA 6*. Herder.

Wulf, C. M. (2005). *Freiheit und Grenze: Edith Steins Anthropologie und ihre erkenntnistheoretischen Implikationen*. Patris.

Wulf, C. M. (2010). *Der Mensch, ein Phänomen: Eine phänomenologische, theologische und ethische Anthropologie*. Patris.
Wulf, C. M. (2021). *Psychotrauma: When the I shatters. Insight in Psychotrauma for Victims and their Accompaniers*. LIT.

Mariéle Wulf, Dr. phil., Dr. theol. habil., MA in education, was professor of moral theology in Tilburg/NL and Professor of Safeguarding in Ottawa/CA. She runs a praxis of pastoral counselling in St. Gallen/CH. Her monographs focus on philosophical-phenomenological anthropology/human dignity, moral-theological subjects, epistemology, psychotrauma and narcissism (in the context of professional ethics). She is an international expert on Edith Stein and her newest research focusses on physical, psychological, ideological, spiritual and structural abuse and how to fight it through safeguarding.

Chapter 11
Rethinking Edmund Husserl's and Edith Stein's Philosophical Anthropology and Metaphysics

Angela Ales Bello

Abstract This paper explores Edmund Husserl's and Edith Stein's analyses of anthropological and metaphysical questions. I begin by examining the meaning of the phenomenological method in Husserl and Stein, underscoring their novel approach to the question of knowledge of the human being, the world, and God. The essay is divided into four parts: two are dedicated to Husserl's and Stein's investigation of the human being, and the remaining two to ensuant metaphysical themes. Concerning the former, we discover a significant connection between the two thinkers; and in the case of the latter, we uncover a metaphysical approach to reality—even present in Husserl—as well as an original development of it in Stein's analysis informed by medieval philosophy and theology.

Keywords Phenomenological method · Gnoseology · Anthropology · Metaphysical investigation of human being · Metaphysical approach to God

The two philosophers who together form the subject of this chapter are linked by a unique teacher-student relationship that unfolds in the history of philosophy as two distinct but related streams of thought, which ultimately render their original theoretical contributions important. Edmund Husserl provides Western philosophy with a new point of view, one which does not simply generate a crisis for all that was understood to be valid in philosophy; rather, it clarifies certain aspects that, in his view, were already present in philosophy but not fully justified. Edith Stein understands the novelty of Husserl's approach, and she helps situate Husserlian phenomenology into the history of Western thought by retracing its essential tenets while bringing it into discussion with ancient and medieval thought. Here, I propose to chart the originality of the phenomenological method and its results for gnoseology while also discussing its applications in philosophical anthropology and Western metaphysics.

A. Ales Bello (✉)
Lateran University, Vatican City, Vatican City State

M. Lebech (ed.), *Stein and Husserl's Phenomenology of the Social in Ideas II*, Women in the History of Philosophy and Sciences 25,
https://doi.org/10.1007/978-3-032-17339-3_11

11.1 Can One Find a Philosophical-Phenomenological Anthropology in Husserl?

The subheading of this section allows us to ponder whether or not a philosophical anthropology can be found in Husserl's philosophy. To begin to answer this question, we must first understand that the term "anthropology" has many senses. Edith Stein offers us various definitions. She distinguishes between anthropology as a: "natural science," which explores the human being as a being of nature; as "science of the spirit," which—following Dilthey, Rickert, and Windelband—views the human being through its spiritual activities; and as "philosophical anthropology." (Stein, 2004). Husserl also discussed different forms of anthropology in two texts found in the volumes *Aufsätze und Vorträge (1911–1921)* and *Aufsätze und Vorträge (1922–1937)*.

The first text, "*Phänomenologie und Psychologie*" (Husserl, 1987, pp. 82–124),[1] was written in 1917, during an important period in which Husserl was deciding the direction of his phenomenology. The text opens with a peremptory affirmation that the "phenomenon" of the human being, which Husserl intends to investigate, must not be confused with the human being studied by the natural sciences. Phenomenology is a "parallel" science that operates on the eidetic plane and seeks to grasp the *a priori* character of phenomena; that is, the "idea" or essential structure of animal subjectivity, personhood, and human sociality, all of which form the basis of empirical sciences like psychology, anthropology, and other sciences of the spirit. Understanding an essential structure is also important for the natural sciences, as Husserl indicates in *Ideas Pertaining to Pure Phenomenology and Phenomenological Philosophy* (Husserl, 1983, section 26). He analyses what he calls "regional ontologies," which revolve around specific phenomena that are assumed to have an essential structure. Anthropology, in this regard, is a regional ontology. The delineation of regional ontologies corresponds to what I define as the first step of the phenomenological method—that is, the reduction to essence, which is then followed by another step; namely, the phenomenological-transcendental reduction, which I will discuss later.

For Husserl, phenomenological inquiry represents a third factor alongside the natural and spiritual sciences; phenomenology is a new science [*eine vollige neue Wissenschaft*] of a philosophical nature. In order to achieve its particular epistemological end, it must bracket all existing scientific knowledge, as Husserl stresses in numerous places of his work. For example, in sect. 56 of *Ideas I*, Husserl writes: "From the standpoint of the phenomenological science which we propose to establish, that signifies [considering] in addition which sciences it might draw from without violating its pure sense, which [sciences] it might and might not use as given beforehand, which, hence, need 'parenthesizing.' Because of its peculiar essence as a science of 'origins,' methodological questions of that sort, which are far removed from any naive ('dogmatic') science, must be considered carefully by phenomenology." All natural and spiritual sciences are bracketed, including those

[1] Stein's elaboration of this paper for publication is published in Stein, 2014, pp. 195–230.

that refer to the human. How, then, does Husserl propose to carry out his research? He adds that what cannot be bracketed is the "cognitive structure of the I," understood in its purity.

Husserl returns to the same argument some 10 years later, when his method has been fully configured, in a text from the 1930s, titled *Phänomenologie und Anthropologie*. The text comes from lectures given at the Kant Societies of Frankfurt, Berlin, and Halle in 1931. In them, Husserl explores the relation between phenomenology and anthropology (Husserl, 1989a). In the beginning of his essay, he recognizes that one finds in the younger generation of philosophers an interest in philosophical anthropology, especially in the "philosophy of life" of Wilhelm Dilthey, whom Husserl criticized in his famous essay *Philosophy as a Rigorous Science* (1910–11). Husserl accuses these philosophers of relativism and notes that the influence of Dilthey is also present in young phenomenologists. One wonders whether Husserl is here recalling Heidegger's discussion of Dilthey in *Being and Time* (Heidegger, 1972, § 77). This claim could be confirmed by Husserl's affirmation that, while phenomenology was born in opposition to anthropologism and psychologism, it seems that a phenomenological philosophy must be built on a conception of "human Dasein" (*menschlichen Dasein*),[2] as was the case with Heidegger's project. But this project is a form of anthropologism because it cannot undergo the phenomenological reduction. In this respect, then, we can understand the differences between Heidegger and Husserl as well the rupture of their human and intellectual relationship.

Interestingly, the charge of anthropologism had already been made by Heidegger against Husserl. In sect. 10 of the first part of *Being and Time*, Heidegger groups together Scheler and Husserl under one criticism, accusing them of not treating the being of the human person as such: "Sie stellen die Frage nach dem 'Person*sein*' selbst nicht mehr" (Heidegger, 1972, 47), although they do offer a more radical form of personalism when compared to other personalist thinkers. What do Husserl and Heidegger mean by anthropologism? The former claims that if we begin with an analysis of *menschlichen Dasein* and if we do not execute a radical *epoché*, we remain on the naturalistic level; the latter, meanwhile, maintains that the results one achieves through the Husserlian phenomenological reductions are linked to a traditional view, against which Heidegger wishes to posit an existential analytic.

In the lecture text we are discussing, Husserl does not share with some others the improper sense and use of phenomenology, but his discussion covers philosophical anthropology, which he attacks and situates as originating in Dilthey's philosophy, with anthropologism and psychologism, whose "isms" indicate an arbitrary absolutization and, therefore, a deviation from the correct method of investigation. Husserl attacks, on one hand, the existential analytic, and on the other—as we saw in *Philosophy as a Rigorous Science*—the temptation of handing over the study of the human being to psychology.[3]

---

[2] "Die Phänomenologische Philosophie soll völlig neu von menschlichen Dasein her aufgebaut werden" (Husserl, 1989a, p. 164).

[3] I treat the relation between phenomenology and psychology in Husserl and Edith Stein in Ales Bello, 2016.

In contrast to anthropologistic and psychologistic views of philosophy, Husserl proposes a field of inquiry that has the transcendental dimension as its core and center. And here, he describes and uses what I call the second step of the method: more precisely, the transcendental reduction. Given the context in which the aforementioned lecture was given—Husserl was addressing Kantian societies—it could be assumed that the use of the term "transcendental" leads to an openness toward Kant's position. But Husserl establishes his distance from the latter; referring, rather, to Descartes, with respect to whom he emphasizes his intellectual proximity (and, at the same time, distance). As is well known, Husserl's relationship with Descartes's philosophy has been constant since the first volume of the *Ideas*, ultimately culminating in the *Cartesian Meditations* (Husserl, 1977). It is also important to note that Descartes is treated again in the Vienna and Prague lectures of 1934, which would come to constitute the core of *The Crisis of the European Sciences and Transcendental Phenomenology* (Husserl, 1970), to mention only a few of Husserl's best-known, published works.

Descartes is regarded as the initiator of the philosophical turn that marks Modern thought; his ideas led to the establishment of a new starting point for philosophy in subjectivity (or, rather, in the egological sphere). However, there are numerous distinctions proposed by Husserl with respect to the Cartesian analysis of the ego. Though Husserl shares the same starting point with Descartes, neither the procedure Husserl follows to reach it, nor the results obtained by the French philosopher are accepted by him. Husserl makes a profound distinction between his position and Cartesian doubt which, by putting the existence of the world into question, surpasses what lies within the realm of human possibilities: the world cannot be denied; it is so present in its existence that it cannot be doubted. The most that can possibly be done is to place it in parentheses or bracket it, and Husserl, as a mathematician as well as a philosopher, reminds the mathematician and philosopher Descartes that everything that is placed in parentheses "lives" in the brackets (Husserl, 83, section 31). The *epoché*, therefore, is only a change of attitude marked by turning away from the naivety of the natural acceptance of everything that comes our way, including the already-established sciences; it leads to the change of perspective that characterizes the theoretical approach of Western philosophy. And what remains is not the human being in its solitude, but the solitude of the transcendental ego immediately filled by the presence of the alter-ego, as we are reminded in the Fifth Cartesian Meditation.

So, what is the transcendental ego? According to Husserl, it overcomes the radical objection that claims that the ego—that is, this human being (*Mensch*)—carries out the method of taking a transcendental position and, in this way, returns to its pure ego. In this case, however, the ego is nothing more than an abstract level of the concrete human being, its pure spirit, since it is abstract with respect to the body. This was Descartes' position. According to Husserl, however, those who maintain this view fall back into a naive and natural attitude, thereby wrongly demonstrating that thinking moves on the terrain of a pregiven world rather than within the domain of the *epoché*: to consider oneself a human being, this is what supposedly grounds the presupposition of the validity of the world. Through the *epoché*, the bracketing of all the knowledge already given and restarting one's investigation, it becomes clear that the ego is the one in whose life the apperception of human being is kept within the universal apperception of the sense of being of the world (Husserl, 1989a, p. 164).

The crucial question arises, then, of whether the reduction to the ego eliminates *Mensch* as *Mensch in der Welt*, the human being as being in the world, to use Heideggerian terminology. Husserl hastens to point out in this lecture that the world remains a fundamental theme and that it is not eliminated, but subtracted, from the "naivety" of everyday knowledge. In this text, he is more committed to ensuring the possibility of justifying the correlation between the world and the subject than to analyzing the structure of the human being—which emerges, in any case, precisely from the deepening of the investigation of the transcendental dimension. This structure had already been analyzed in *Ideas II*, which Edith Stein transcribed from Husserl's Gabelsberger shorthand. Among his published works, this is in fact the most complete text delineating a philosophical anthropology. It is a new way of performing transcendental analysis that allows us to identify the flow of experiences (*Erlebnisse*) of which we have "awareness" and that can be understood as a tool for understanding all the dimensions of the subject. Herein lies the novelty of phenomenological analysis with respect to traditional gnoseological accounts of Western philosophy; in fact, described essentially, these experiences refer to the "real" structures of the human being. Hence, *Ideas I* and *II* are intimately connected and must be read as constantly referring to one another. If the first volume aims at clarifying the phenomenological method while delineating the domain proper to phenomenology (namely, the transcendental dimension, understood as locus that can reveal the sense of the *quoad nos*), then the second volume seeks to bring to light the constitution of material nature, to which belongs the belong the body, that is, animal nature, which is intimately linked with a psychic reality and a spiritual world. The personal I belongs to this spiritual world.

I maintain this interpretation because I could find the same main points regarding the structure of the human being in the original Husserlian texts which are to be published in the new edition of *Ideas II* (Husserl, 2025). The Husserlian texts are not elaborated, but all the same we find in them the description of *Leib, Seele* and *Geist,* discovered and justified through lived experiences (*Erlebnisse).* I think that, in reporting Husserl's analyses, Edith Stein was faithful to his texts, even if she organised them according to her own reading; but we must remember that she asked Husserl insistently to oversee her transcription, and he had no time to do it.

It is interesting to note how, after having bracketed all the traditional views of the human being, the phenomenological analysis of it is approached in an original way: the method is not deductive but ostensive, recovering and confirming the longstanding perspective of human being's tripartition in body, psyche and spirit that we first find in Saint Paul's writings. The notion of *Leib*, the living body, is exquisitely phenomenological. The very structure of the human being comes to evidence from the examination of the living body, which is not in itself a starting point; rather, its characteristics are traced by moving from the presence of "perception" as an experience of which one is aware. If perceptual apprehension presupposes the contents of sensation, which play a necessary role for the constitution of *schemata* and thus for the constitution of the appearings of real things themselves, this implies "the fact that in all perception and in perceptual exhibition (experience), the Body is involved as freely moved sense organ, as freely moved totality of sense organs, *and hence there is also given the fact that, on this original foundation, all that is thingly-real in the surrounding world of the Ego has its relation to the Body.*" (Husserl, 1989b, p. 61).

Once the living body has been identified in *Ideas II* (from which I took the foregoing quotation), one would not expect a resumption of the treatment of the theme of consciousness lived through the pure ego, understood as the essential reference point of the current of experiences, which Husserl had already described in the first volume. But this recovery serves as a reminder that the description of consciousness' essential features is possible thanks to the capacity of human beings discovered through reflection. Hence, "it is by no means whatsoever something mysterious or mystical. I take myself as the pure Ego insofar as I take myself purely as that which, in perception, is directed to the perceived, in knowing to the known, in fantasizing to the fantasized, in logical thinking to the thought. [...] To speak more precisely the pure Ego is farther related to Objects in very different modes, according to the type [...] accomplished." (Husserl, 1989b, pp. 103–4). These acts, grasped in their universality, can be of desire, love, hatred, attraction and repulsion, decision in an action, a *fiat*, willing, and even a theoretical act, including delineating a thematic context, establishing relations, positing a subject and a predicate, and drawing consequences. In addition, a *sui generis* transcendental structure is delineated that allows us to move on to an investigation of the human being as nature—the living body as a bearer of localized sensations or as crossed by the sensations of pleasure, pain, well-being or discomfort, which constitute the material basis of nature. Husserl defines this basis as hyletic; that is, it serves as the basis for the constitution of values. This originary layer, therefore, permits intentional functions to be connected, and matter can take on spiritual functions.

Through the layer of "real" qualities (hyletic basis), insofar as they are constituted by virtue of a relationship with real circumstances in the realm of reality, the living body is intertwined with the psyche: "Soul and psychic Ego 'have' a Body; there exists a material thing, of a certain nature, which is not merely material thing, but is a Body, i.e. a material thing which, as localization field for sensations and for stirring of feelings, as complex of sense organs, and as phenomenal partner and counter-part of all perceptions of things, [...] makes up a fundamental component of the real givenness of the soul and the Ego." (Husserl, 1989b, pp. 165).

There also exists a qualitatively distinct dimension, different from the material thing; namely, psyche. The living body is intertwined with psyche. Some Husserlian manuscripts, further exploring the dimension of psyche, analyze the realm of instincts, which indicates a continuity with the animal world; however, this domain is distinguished from the one marked by intentional and spiritual functions. (Ales Bello, 2000). We saw earlier that the pure ego also includes experiences that are different from tensions, impulses, and reactions; these are the voluntary, evaluative and theoretical acts that characterize the human person. In this way, one enters the life of the spirit, which is in no way "determined," but rather "motivated." It is the seat of free acts and rational standpoints. Passivity and activity intertwine, but activity characterises the "awake" or "wakeful" human being, who is ethically and theoretically awake.

We proceed, therefore, to the definition of the person, who has representations, who feels, evaluates, pursues something, acts; and in each of these personal acts, s/he stands in relation to some end, with the surrounding objects of her/his world. Taking a personalistic attitude toward the surrounding world means adopting an evaluative and ethical attitude. This, far from being artificial, is the real "natural" attitude. It is one of the few places where the term "natural" is used by Husserl, for he usually excludes it because he attaches it to the use that Positivism makes of it in the context of naturalism.

Certainly, the Husserlian position is not a substantialist position. All this is far from his mental horizon for a series of reasons: his scientific training; his personal approach to philosophy, which is without ties to a particular current of thought; his belonging to the Protestant cultural world in conflict with medieval philosophy and for its rejection of the rationalistic metaphysics of the Modern Age. However, as is well known, he was introduced to philosophy through Brentano, a former Catholic priest, a philosopher who was heir to ancient and medieval thought, an Austrian who remained faithful to the Church of Rome, and a supporter of a rigorous philosophy open to the new demands of psychology.[4]

Edith Stein affirms that the foregoing analyses offered by Husserl are not secondary for understanding the "essential" description of the human being—an essentiality that does not immediately unfold within a metaphysical framework, but which nonetheless allows for the delineation of the human being, with its own characteristics that allow it to affirm its excess with respect to what is "natural." The result, after the *epoché* of all the interpretations already given, is—in truth—the recovery of the Greek-Christian Western tradition: phenomenology, understood as a phenomenological philosophy, arrives at a description that strengthens the tradition, but does so by following a new path.

In Husserlian philosophical anthropology (we can, in fact, now use this expression: an anthropology that is outlined in a phenomenological framework and therefore, as I have already indicated, is ostensive and not speculative), much space is dedicated to the ethical-religious dimension. Undoubtedly, the gnoseological aspect in Husserl's research prevails, and it certainly is important as a way to solve all problems; but this in no way means that he has set aside his private research into the "ultimate and supreme problems," which he defines as "metaphysical" in one of his most important works, the *Cartesian Meditations*. Ethical-religious questions are linked to the question of God and are addressed in numerous places in his work, as I have shown elsewhere. It seems important to me to mention these aspects of his anthropology, from which emerge a human being examined in all its potential, that is not simply reduced to some of its dimensions: the human being is open to others. Think of the Husserlian

[4] For more information on this argument, see Ales Bello, 2016.

analyses of intersubjectivity. Here, we can also add the Husserlian idea of the interpersonality and openness to the Other as the ultimate justification of the existence of all human beings. I will return to this topic later.

The Husserlian position on anthropology is corroborated and confirmed, as indicated above, by the very same objections that Heidegger makes to Husserl and Scheler, both of whom are Heidegger's target of critique. In §10 of *Being and Time*, we read that the interpretation of the human being as a corporeal-animated-spiritual unity, the Husserlian and Schelerian one, is absolutely insufficient, because it is not possible to conceive of this being by joining together modes of being such as the body, the soul and the spirit, which are assumed to be totally indeterminate in their being. Furthermore, an attempt at an ontological investigation of this kind would be forced to suppose an idea of the being of the whole, and this is linked to the orientation of Greek-Christian anthropology, which has conjugated the definition of human being as an *animal rationale* with the being and essence of the theological order. Here, we are not interested in developing what Heidegger criticizes; rather, I wish to emphasize that what for him was an accusation inevitably reveals itself as the fundamental sense of the anthropology proposed by Husserl.

I would like to conclude the brief synthesis of what I call Husserl's philosophical-phenomenological anthropology by underlining that it is possible to answer affirmatively the question posed at the beginning of this paper. Even if Husserl did not accept the definition of "philosophical anthropology" to indicate the path taken and the results achieved, he showed—through an analysis that starts from the human subject and returns to the human subject itself, which presupposes, as he himself indicates, the paradox of this entity—that the human being is the only existent capable of simultaneously being the subject and object of phenomenological investigation, in a non-assertive, but as ostensive a way as possible; he shows the complexity of its structure.

11.2 The Acceptance of the Phenomenological Method and the Analysis of the Human Person in Edith Stein

Edith Stein, *contra* Husserl, accepts Scheler's term "philosophical anthropology," though she generally follows Husserl's description of the human person in her early phenomenological works. Here, I would like to demonstrate the centrality of the human being for Stein's phenomenology. In fact, the human being is the primary object of investigation, as evidenced in her first major work on empathy, *On the Problem of Empathy* (Stein, 1989). In this work, Stein examines how one comes to grasp the other in relation to one's own self-knowledge, which in turn leads to the discovery of a shared human constitution. Also important here is the analysis of spirit, which leads Stein to articulate the unique specificity of human beings in relation to one another and other beings.

Stein's inquiry into the essence of the human being is characterized by the attention she gives to the reality that surrounds us. The phenomena that we encounter, understood as "manifestations" that are offered to us, present themselves as having a traceable essence. This vision of a delineable essence could be viewed as optimistic, though it is subject to confirmation and verification through the results of the research; an essence is not presupposed *a priori*, and so here the human being is able, to some extent, to grasp the "sense of things." The term "thing" has a plurality of meanings; it does not refer only to the thing in the physical sense. In German, the word for thing is *Ding*, but also *Sache*, which can be translated as "the question" or "the argument" or "the fact that must be understood, and that lies between the facts." In the phenomena he encounters, Husserl privileges the human subject. Husserl and Stein invite us to explore the phenomena we encounter, not only in our singularity but also in our common humanity that we find in ourselves and in others.

In the context of phenomenological analysis, it is always necessary to underline the importance of the moment of "discovery", to highlight the fact that nothing is being taken for granted or presupposed. It is necessary to bracket all received definitions of human being, both from the viewpoint of the philosophical tradition and from that of the natural and human sciences. As we saw, Husserl preferred the widely used expression "the sciences of the spirit," which was prevalent in the philosophical milieu of his day. The reasonableness of the use of the *epoché*, of the suspension of judgment, of reduction, lies in the fact that no solution, can be presented as absolute, as definitive and indubitable; rather, each essential description must be thought of as always raising questions and problems. Starting over has, therefore, the sense of sifting through what is said to see whether it is said because that is how the thing under consideration is shown, in its essence—that is, in its structural characteristics, highlighted through a purified gaze in such a way that we can intuitively grasp what is given, not by stopping at the surface, but by proceeding to carry out the work of excavation. This operation guarantees the result.

Inspired by the investigations of her teacher Husserl—in particular, those of *Ideas I* (1913)—Edith Stein chose as the theme of her dissertation (which she defended in 1917 at the University of Freiburg) the concept of knowledge of the other through a deepening of the meaning of a particular act that we constantly perform. The name for this is *Einfühlung*, which we can translate with the term entropathy[5] or empathy. As noted earlier, in his investigative work on subjectivity, Husserl explores precisely the terrain that Kant had already identified and defined with the term transcendental. However, this examination leads Husserl to results different than Kant's, insofar as Husserl discovers unique acts of consciousness that we live, that are "lived" by us. These acts characterize our "interiority" (the latter expression is not Husserlian; I use it only to make people understand which sphere is being examined). Experiences or acts are structural elements which we all possess, while their contents can be extremely subjective, relative, and changeable.

[5] I use this word, which in Italian (*entropatia*) translates the Husserlian word *Einfühlung*, in order to distinguish what empathy means in phenomenology from what it means in psychology (e.g., Kohut) or in common language (empathy as sympathy).

At the beginning of her dissertation on entropathy, Stein takes up the fundamental aspects of the Husserlian method by distinguishing the essential characteristics of the various experiences she explores. To understand her trajectory, let us return to the path followed by Husserl. Husserl begins with an act that he considers very important because it places us in relation with the outside world; namely, the act of perception. By examining it in itself, in an essential way, and by trying to understand what it consists of as an act, we understand that it puts us in direct, immediate contact with reality; it is an act that we can all perform under normal conditions, but the content of perception is changeable, because it depends on the circumstances and on the perceived objects. In carrying out this analysis, however, we have isolated it, because we live it together with a series of acts of which we are aware; it is the awareness of these acts that accompanies us. Stein describes it as a light that illuminates experiences, but the experience of this light does not mean that we have thematized these acts. To do this, we need a further and particular experience—namely, reflection. It allows us to place ourselves in the acts we live in another attitude, which allows us to grasp their meaning.

If we analyze what happens in ourselves, we realize that not only the act of perceiving is present, but also the acts of remembering, imagining, fantasizing, and thinking; we can continue identifying a multiplicity of acts or lived experiences—a task Husserl executed throughout his entire body of research. Consciousness also records experiences relating to impulses, instincts, tensions, as well as ones related to the psychic sphere. We also find lived experiences of decision, of the willed taking of conscious positions, which in turn qualitatively constitute another group of acts that stem from the realm of spirit. Entropathy, as Husserl observes, can be included among the experiences that can be identified through the analysis of lived experience.

Examining this peculiar act in which the knowledge of the other is considered entropathic knowledge, Edith Stein asks what is grasped of the other and through what instrument is it grasped. The result of the analysis with respect to these two questions constitutes a further confirmation of the complexity of the human being, which I have mentioned above. Where Husserl conducted his investigation by starting mainly from the complexity and stratification of the structure of human being as such, his student instead first derives the fundamental connotations of anthropology precisely through the analysis of entropathy.

Stein discovers the human being as a psychophysical individual. On the one hand, corporeity is involved as a necessary means for knowledge itself. On the other hand, corporeity is: animated; living (*Leib*); a seat of affects and drives; a locus of thoughts, reasoning, decisions, and volitions. Are these acts qualitatively the same, can they be placed on the same level without distinction; or do they refer to different gradations that manifest different values? In other words: what is the genesis of the traditional interpretation of "soul" and "body"? And, in what sense is the human being not reducible to corporeity, even though this is the indispensable dimension of life as it is given; that is, as we observe it?

It is in the territory of conscious lived experiences that body and soul come to be identified, as has already been noted. They constitute specific configurations of human being, consolidated by a long tradition and present, even if in different forms, in all cultures. The pain or joy that the other experiences must be grasped first through a perception linked to the organs of sensation; but what one sees on the other's face reflects a depth of life that resists identification to or with any physical object that is seen and touched. Here, a dimension is discovered that is the locus of affections, drives, and emotions, and which we can denote with the term psyche (*Psyche*).

One might ask—and Edith Stein re-poses this question in the second decade of the twentieth century, a question already asked by the "master"—is there a discipline that scientifically investigates and truly shows the psyche as it is without reducing it to its mechanistic structure (action and reaction, association, and so on, as the then-nascent psychology seemed to do), ultimately appropriating the method of the natural sciences? From this question arises the long and complex essay on psychology and the sciences of the spirit, published in 1922 in the "Yearbook" of phenomenology edited by Husserl and translated as *The Philosophy of Psychology and the Humanities* (Stein, 2000a).

The intent of the text is not to eliminate psychology in favor of a purely philosophical approach, but rather to show the insufficiency of the then-current view of psychology and of the new human or social sciences, established in the second half of the nineteenth century; in particular, sociology. The claim of insufficiency revolves around the lack of a deep understanding of the human being and the justification of these social sciences' constitution as science. For Stein, the social sciences first require a preliminary investigation that clarifies the underlying moments or aspects on which they rest. For example, psychology cannot help but ask itself what the psyche is, and must also ask itself whether corporeity and psyche are sufficient to understand the human being in its entirety. These questions require philosophical reflection.

Through the analysis of lived experiences, Stein highlights the existence of a psychic sphere characterized by a "vital force," by psychic energy, which plays a fundamental role for the human being's very survival and ability to act. Yes, it is structured according to a sort of causality, as relations among psychic events can be considered deterministic to a certain extent, but is this very causality the only source of direction for the human being? The psyche's states of feeling fresh or fatigued, for example, are always and exclusively the basis of our actions (and therefore of our decisions), but are there also choices that refer us to "reasons" that reveal a different source of decision? It is here that the sphere of the spirit (*Geist*) becomes manifest.

Entropathy had already been identified as the tool for understanding that which is present in the life of the other, and not only in relation to his/her feelings or emotions, but also in relation to his/her inner world of decisions and volitions; in relation to meaning or sense in his/her world, to creativity. If the sciences of the "spirit" exist, they in fact concern that which is connected with the sphere of cultural, artistic, and political production. From a philosophical point of view, the presence in the human being of the dimension of the spirit allows us to affirm that the psycho-physical reality that brings us closer to the animal world is overcome thanks to an element that defines the human being as a "person."

The aforementioned position will remain fundamental for Stein, even when her research turns to ancient and medieval thought. Indeed, it will help confirm her later phenomenological analyses, especially in her works *The Structure of the Human Person* (1932) (Stein, 2004) and *Finite and Eternal Being* (1936) (Stein, 2002). In the latter, for example, Stein pauses to investigate finite beings: here, among them, we find the human being, and by examining it, one finds that "its body" is a material body. However, one simultaneously notes that the human being is an organism that takes shape and acts from within. The human being is not only a physical organism, for it is an animated, living being who—in a particular way—is perceptually open, is open to itself and its surrounding world. The human being is discovered to be a spiritual being who is cognitively open to itself and to the larger reality; it can freely shape itself and reality, and as such is described as a "person" by Stein. The term "person" is not only present in the analyses of Husserl and Stein, but also refers—for both philosophers—to the same essential characteristics of the human being, especially to its spiritual component.

11.3 Husserl and Metaphysical Questions

As noted earlier, Husserl speaks of ethical-religious and metaphysical problems as the ultimate problems. The term "metaphysics" has many meanings in Husserl: we pause briefly to note that he distinguished a "historically degenerate metaphysics" and a metaphysics that was originally founded as *first philosophy* that by "no means conforms to the sense with which *first philosophy* was instituted originally." (Husserl, 1977, p. 139). The former refers to an investigation rooted in an "excess of speculation," whereas the latter refers to the authentic spirit of Greek philosophy, as Aristotle largely understood it. However, the reference here to first philosophy is not only linked with Plato or Aristotle but also with Leibniz's metaphysics, in particular with the notion of the "monad." According to Husserl, my ego—understood as a monad—"can be a world experiencing ego only by being in communion with others like himself: a member of a community which is given orientedly starting from himself. In that the objective world of experience shows itself consistently, the other monads show themselves to be existent." (Husserl, 1977, p. 139).

For Husserl, there is only one objective world, which is identifiable through what he calls transcendental metaphysics. This demonstrates how, departing from the subject, one can understand the constitution of an objectively existing world. I have called this position of Husserl's, which places together terms traditionally considered opposed to one another such as the transcendental ego and the objective existence of the world, "transcendental realism." (Ales Bello, 2015). Husserl maintains that it is precisely the passage to otherness, always through a lived body, that guarantees an existing objective world; this possibility is the foundation of "transcendental" realism: "But what about the other egos, who surely are not a mere intending and intended *in me*, merely synthetic unities of possible verification *in me*, but according to their sense, precisely *others?* Have we not therefore done transcendental realism an injustice? The doctrine may lack a phenomenological foundation, but essentially it is right in the end, since it looks for a path from the immanence of the ego to the transcendency of the Other." (Ales Bello, 2015, section 42, p. 88). Husserl sees his new metaphysical point of view as valid because it is not a question of a "metaphysical" construction linked only to a tradition, but of a position justified on the basis of intuition, which tends to "explicate the sense this world has for us all, prior to any philosophizing, and obviously gets solely from our experience—a sense which philosophy can cover but not *alter*." (Ales Bello, 2015, p. 151). Following this line of analysis, which tends to emphasize the sense of an existing reality in which we find ourselves immersed and which is defined by Husserl as "transcendental metaphysics" of immanent realities (that is, the I and the world), Husserl then moves on to analyze the metaphysics of absolute transcendence.

I have identified six "argumentative ways" in Husserl's writings that lead to the metaphysical question of God: the "objective" way of finality that recalls the fifth teleological way of Thomas Aquinas; the subjective way that begins from Augustine and Anselm; the intersubjective or Leibnizian way, which leads to the Ultimate Monad; the way through hyletics, which is always a teleological way that starts from the deep dimensions of reality that are pre-reflective and precategorial and which manifest the presence of an order of things; the way through ethics; and perhaps a "mystical" way that deepens the presence of God in our interiority (Ales Bello, 2005).

I am unable here to discuss at length these ways, but I would like to pause briefly on the path that leads to the Highest or Ultimate Monad to continue to develop the idea of the monadic metaphysical constitution of the self and the world. If we are the starting point of phenomenological investigation, we constitute a sort of absolute *quoad nos*. Husserl writes in a text from 1922: "From this absolute there must be, then, a path that leads to the ultimate [or highest] absolute in another sense; from this system of 'substances' 'in the true sense' (as beings that do not presuppose the being that constitutes them) there must exist a way to absolute substance in the ultimate sense."[6] It has been said above that Husserl's position is not substantialist; in this text, the term "substance" appears, which is probably a simple taking up of the Leibnizian usage of defining monads as substances. Husserl continues to make explicit reference to the Ultimate Monad as the point of arrival on the path of intersubjectivity to God. This reference is present in particular in another text from 1922, titled "Possibility of the Fusion of Monads: Possibility of an Ultimate (Divine) Monad."[7]

We cannot treat here the ethical-religious themes present in Husserl's work. I only observe that Husserl's treatment of ethical issues is, in some cases, connected to religious themes and, as I said above, that there is also an ethical way that leads to God. Husserl does not believe that there exists a blind destiny; on the contrary, God rules the world and directs it towards absolute values that appeal to freedom and the human will, which is supported by grace. It is this directedness that fortifies

[6] Edmund Husserl, "*Die transzendentale Alter Ego gegenüber der Transzendenz der fremden Subjektivität des Dinges. Absolute Monadologie als Erweiterung der transzendentalen Egologie. Absolute Weltinterpretation*, (Januar/Februar 1922)", (Husserl, 1973, 266).

[7] Edmund Husserl, "*Möglichkeit der Verschmelzung der Monaden. Möglichkeit einer (göttlichen) Übermonade (1922), Beilage XLI*" (Husserl, 1973, 300–302).

Husserl's belief in God's existence: "Faith is the absolute and highest requirement." (Husserl, 1924, p. 16). It is no coincidence that this is a manuscript not intended for publication. Here, the metaphysical problem concerning the Absolute is intertwined with the ethical-religious one. In this case, Husserl, more than simply proceeding with a rigorous argument, exposes his deepest and most personal thoughts.

11.4 Stein on Metaphysical Questions

Stein passes seamlessly from a phenomenological analysis to a metaphysical position. The subject is not only a pure I, psychic I, and spiritual I, but also "substance." (Stein, 2001, Chap. 5. 2). Its separate being is graspable by the subject's consciousness and it is different from anything else. It alone can call itself I. However, it should be remarked that Stein remains on phenomenological ground when she takes up the themes of intentionality, temporality, the spiritual life as an intellectual life, and the connection between interiority and exteriority (Stein, 2001, Chap. 5. Sect. 4, 5 and 6). The same move described above happens for the moral life, in which the other power of the spirit emerges—that is, the will. Human beings open themselves up to the external world, with their whole person, which includes corporeity, psyche (pleasure and displeasure), feelings (love and hate), and emotional tonalities. Everything converges from the subjective side toward the formation of character, which is also linked to the characteristics of individuality. The complex, layered map of the inner life is completed by the presence of a nucleus or core, already identified in Stein's phenomenological writings and now defined by her as the soul of the soul—a deep point of convergence of the human being and a place of opening towards transcendence, constituted by others and God.

Drawing on Thomas Aquinas, the spirit (*Geist*) increasingly reveals itself as the seat of intellectual and voluntary operations in the double cognitive and moral sense. All this had already been said by Husserl and by Stein herself, but here one must not forget that this comes to be ultimately based on a metaphysical ground. And the nucleus or core (*Kern*) of the person, from the point of view of ontology, acquires a further determination in the wake of the analyses already carried out in Stein's earlier phenomenological works (especially in *Philosophy of Psychology and the Humanities* and *The Structure of the Human Person*).

As for the question of knowing and demonstrating the existence of God, Stein's attitude is unique. She repeats several times that "proofs" or "demonstrations" do not help find faith; they validate faith, but they do not lead to faith. The believer—that is, the one who already has faith, understood as a knowledge of God—allows him- or herself to be led by various proofs or tests. The non-believer remains on the borders: "The thinker who applies the rules of natural knowledge shrinks back again and again from the leaps over the abyss." (Stein, 2002, p. 110). This does not mean that faith is an irrational fact; on the contrary, Stein maintains that a *fides* that accepts Revelation can—indeed, must—help natural reason acquire the knowledge of the truth. In this sense, reason, enlightened by faith, can become "supernatural reason"—an expression that Stein uses only in her comparative essay on Husserl and Saint Thomas (Stein 2000b) but which, in an important sense, she always keeps present, because it serves to delimit the boundaries of human reason while accepting the possibility of its being strengthened through faith. If philosophy wants to reach the truth, it must recognize the different sources of truth, including Revelation, which needs *fides*—understood as natural knowledge—to be accepted.

The foregoing relational circularity of faith and reason is kept in mind by Stein when in *Finite and Eternal Being* she reflects on the meaning of "Eternal Being." (Stein, 2002, Chap. 3, sect. 12). Drawing on two scriptural texts, Stein observes that they grant us deeper understanding than a purely intellectual pursuit can achieve. Specifically, she refers to the beginning of John's Gospel ("In the beginning was the *Logos*") and to the passage from the Letter to the Colossians 1:17, in which Paul writes: "He is before all things and in Him all things hold together." And she adds that perhaps the philosophical meaning of the *Logos*, to which we her readers are referred, can help us understand a deeper theological meaning. Hence, for Stein, revealed truth can solve philosophical difficulties. Let us briefly proceed to a philosophical clarification of the foregoing two texts.

For Stein, the two texts refer to the Trinitarian vision of God. In particular, *ev arche ev o logos* has the following meaning: the *Logos* (the divine sense or essence) was included in the First Entity, in the Father was the Son; that is, the "sense" of the original actuality. Therefore, generation refers to the position of the essence in the new personal actuality of the Son—which, however, is not a position outside the Father's original actuality. In distinguishing the essential being from the actual-real being, though, there is a risk of separating it in thought, and this cannot be done. Stein emphasizes the validity of the position of St. Thomas, who insists on the inseparability of essence from actual-real being in the First Being, in whom Being is distinguished from any other finite being in which one can make the foregoing distinction.

Stein observes that, since the first Being has being by essence, it is impossible even to think of it without being. But we must ask ourselves what this "thinking" is. Indeed, if we were able to think truly in a fulfilling way—and here Stein uses the phenomenological language of the total fulfillment of an empty intention—we would be faced with a profound understanding of what Anselm of Bec proposed as "proof" for the existence of God. However, she reflects further on this question, arguing that, if we could "truly" grasp the coincidence of being and essence, we would not be faced with a "proof" and, therefore, from a logical point of view, a "conclusion" or deduction; rather, we would have shaped the "originary thought" ("eine Umformung des ursprüngliches Gedankens"), which we come to grasp in our minds. And she adds that the accuracy of this shaping is not contested even by Thomas Aquinas, who rejected the ontological proof: "St. Thomas, who, as is well known, rejects the proof of St. Anselm, does not deny the validity of this transmutation." (Stein, 2002, p. 109).

Even Thomas, after all, affirms that the proposition "God exists" is self-evident, but not for us. It is precisely because we fail to truly grasp this evidence that we need a proof starting from the effects. Even in this way, however, we do not arrive at understanding, and here, in the final analysis Stein holds that Augustine is right to say: "Si comprehenderis non est Deus" ["If you were able to understand, then it is not God"]. So too Anselm, who tries to give a form to the originary thought, fails because he does not arrive at full understanding. The profound reason for the failure of all proofs or ways of achieving unreserved fulfillment lies in the unresolvable tension in the human spirit between the finite and infinite. And this also makes us understand, according to Stein, the singular destiny of the ontological proof—namely, that there are always new defenders and adversaries of it. It is therefore not possible to grasp the coincidence of essence and existence through the path of natural knowledge. For those who have faith, it seems impossible that there be no coincidence; but when one tries to understand it intellectually, one is doomed to fail. One is forced to approach God only through finite images, sometimes from the vantage point of essence and sometimes from the vantage point of being.

As is well known, Stein maintains that a full knowledge of God, in accordance with the human being's abilities (which are always limited as s/he is a creature) is possible for the human being only in mystical experience—because God thereby raises the soul, causing us to intuitively grasp the meaning of all things, according to the testimony of Saint Teresa of Ávila (Stein, 2006). Therefore, the mystical experience illuminates self-knowledge and is of great help for understanding both our anthropological structure and God. Metaphysical themes, even though they can be treated independently with respect to faith and mystical vision, can be illuminated and more easily dealt with by these peculiar types of "knowledge."

11.5 Concluding Observations

Husserl and Stein do not shy away from the big questions like: Who am I? What is the most graspable sense of the reality and of ultimate reality? With lesser or greater proximity to the classical and constructive positions of the history of philosophy and in conversation with Modernity (Husserl 2025) or with ancient and medieval thought (Stein), through an original and convincing approach, both philosophers show how we cannot escape these great questions, and how it is possible to give valid answers to them.

According to my interpretation of the works of Edmund Husserl and Edith Stein, I would like to highlight some points that I find particularly significant. First, the theme of anthropology in Husserl. I understand that he did not wish to use this term, which belongs in his time to the natural sciences, because he considered the human being to be linked, rather, to the spiritual sciences. Edith Stein's use of the term "philosophical anthropology" allows us to affirm that it can also be attributed to Husserl's analyses of the human being.

Second, the reprinting of Husserl's texts that make up *Ideas II*, vol. IV/V in the *Husserliana* series, shows that Edith Stein's revision was very faithful to her teacher's intentions. She organized the manuscripts so that they were more readable but did not alter their content.

A third question concerns my proposal to interpret Husserl's phenomenology as "transcendental realism." I observe that Husserl's position goes beyond idealism and realism, as they are currently understood, but the correlation between subject and object, as proposed by him, leads us to consider the relationship between transcendental subjectivity and external reality in its existence. The phenomenological method allows us to better understand this relationship.

The last issue concerns the presence of metaphysics in Husserl, which had been contested by many interpreters and which I have traced in his works and manuscripts, indeed the three great problems of Western philosophy: the self, the world, and God, are addressed by him with great acuity and originality.

References

Ales Bello, A. (2000). *Human world–animal world: An interpretation of instinct in some late Husserlian manuscripts. AHua* vol. 68, 249–253.

Ales Bello, A. (2005). *Edmund Husserl: Pensare Dio Credere in Dio.* Padova: Edizioni Messaggero. English translation: Ales Bello, A. (2009). *The Divine in Husserl and Other Explorations* (A. Calcagno, Trans.). *AHua,* vol. 98. Springer. Spanish translation: Ales Bello, A. 2016). *Edmund Husserl, Pensar Deus, creer Deus* (L. Rabanaque, and M. Calello, Trans.). Editorial Biblios–Paideia Fenomenológica. Portuguese translation: Ales Bello, A. (2016). *Edmund Husserl: Pensar Deus, Creer Deus* (A. Turolo Garcia, M. L. Fernandes and T. Akira, Trans.). Paulus.

Ales Bello, A. (2015). *The Sense of Things: Toward a Phenomenological Realism.* (A. Calcagno, Trans.) *AHua,* vol. 118. Springer.

Ales Bello, A. (2016). *Il senso dell'umano. Fenomenologia, psicologia e psicopatologia.* Castelvecchi.

Heidegger, M. (1972). *Sein und Zeit.* Max Niemeyer Verlag.

Husserl, E. (1924). Ms. A V21, "*Ethisches Leben, Theologie; Wissenschaft, 1924*". (A. Ales Bello, Trans.)

Husserl, E. (1970). *The Crisis of European Sciences and Transcendental Phenomenology* (D. Carr, Trans.). Northwestern University Press.

Husserl, E. (1973). In I. Kern (Ed.), *Zur Phänomenologie der Intersubjektivität II.* Martinus Nijhoff.

Husserl, E. (1977). *Cartesian Meditations* (D. Cairns, Trans.). Martinus Nijhoff.

Husserl, E. (1983). *Ideas Pertaining to a Pure Phenomenology and to a Phenomenological Philosophy. I* (F. Kersten, Trans.). Martinus Nijhoff.

Husserl, E. (1987). Phänomenologie und Psychologie. In T. Nenon & H. R. Sepp (Eds.), *Aufsätze und Vorträge I, Hua* (Vol. 25, pp. 82–124). Martinus Nijhoff.

Husserl, E. (1989a) *Phänomenologie und Anthropologie,* in *Aufträge und Vorsätze* (1922–1937), in *Hua* vol. 27 (T. Nenon & H. R. Sepp, Eds.) Martinus Nijhoff.

Husserl, E. (1989b). *Ideas Pertaining to a Pure Phenomenology and a Phenomenological Philosophy, II* (R. Rojcewicz & A. Schuwer, Trans.). Springer.

Husserl, E. (2025). Ideen zu einer reinen Phänomenologie und phänomenologischen Philosophie, zweites Buch: Phänomenologische Untersuchungen zur Konstitution und Wissenschaftslehre. Die drei Teilentwurfe und Thematische Ausarbeitung sowie das, Nachwort zu meinen Ideen (1908–1930). In D. Fonfara (Ed.), *Hua, IV/V.* Springer.

Stein, E. (1989). *On the Problem of Empathy* (W. Stein, Trans.). CWES III. ICS Publications.

Stein, E. (2000a). *Philosophy of Psychology and the Humanities* (M. C. Baseheart & M. Sawicki, Trans.). CWES VII. ICS Publications.

Stein, E. (2000b). "Husserl and Aquinas: A Comparison", in *Knowledge and Faith*, (W. Redmond, Trans.). CWES VIII. ICS Publications.
Stein, E. (2001). *Potency and Act* (W. Redmond, Trans.). CWES XI. ICS Publications.
Stein, E. (2002). *Finite and Eternal Being: An Attempt at an Ascent to the Meaning of Being* (K. Reinhardt, Trans.). CWES IX. ICS Publications.
Stein, E. (2004). Der Aufbau der menschlichen Person: Vorlesung zur philosophischen Anthropologie. In B. Beckmann-Zöller (Ed.), *ESGA 14*. Herder.
Stein, E. (2006). Die Seelenburg, in endliches und ewiges Sein. Versuch eines Aufstiegs zum Sinn des Seins. In A. U. Müller (Ed.), *ESGA 11/12*. Herder. Anhang II.
Stein, E. (2014). Freiheit und Gnade und weitere Beiträge zu Phänomenologie und Ontologie (1917–1937). In B. Beckmann-Zöller & H. R. Sepp (Eds.), *ESGA 9*. Herder.

Angela Ales Bello is Professor Emerita of History of Contemporary Philosophy at Lateran University in Rome and past Dean of the Faculty of Philosophy. She is the President of "The Italian Center of Phenomenological Researches" (Rome) affiliated to the World Phenomenological Institute, U.S.A., President of "The International Society of Phenomenology of Religion", Rome, Italy, and past Director of the International Research Area dedicated to "Edith Stein and Contemporary Philosophy" at the Lateran University. She is Visiting Professor in the Faculty of Philosophy of the Pontifical University "Antonianum" in Rome, in the Faculty of Psychology of the State University in São Paulo and of the Catholic University in Campinas, Brasil.

Chapter 12
Concluding Observations

Mette Lebech

Abstract These concluding observations attempt to draw together what we can know about the relationship between Stein's and Husserl's phenomenological positions in and around the time of *Ideas II*'s redaction by Stein. They read Stein's *On the Problem of Empathy* as testifying to the presuppositions with which she started her work on *Ideas II*, and her *Contributions to the Philosophical Foundation of Psychology and the Humanities* as stating her critique of what she took Husserl's position to be in the work she put together for him. It is argued that *On the Problem of Empathy* provides the rationale for the structure of the 1952 edition of *Ideas II* and that *Contributions* explains the reasons behind that structure in the form of a critique of the result.

Keywords Husserl · Stein · *Ideas II* · *Beiträge* · Psychic causality · Body · Motivation · Psyche · Empathy · Psychology

12.1 Introduction

To conclude this work, it seems useful to sum up what has been learnt in the process of compiling it. To this end, we shall frame our considerations by exploring two works published by Stein, the compositions of which respectively precede and postdate her work on the manuscripts of *Ideas II*. As these works—*On the Problem of Empathy*, (Stein, 1917, 1989) and *Contributions to the Philosophical Foundation of Psychology and the Humanities*[1] (Stein, 2000, 2010)—indisputably express Stein's considered

[1] Throughout, I alternate between four different ways of designating this work: by its full name, as here, by "*Philosophy of Psychology and the Humanities*", which is the title of the current English translation published in the Collected Works of Edith Stein (CWES) from ICS Publications; by "*Beiträge*", which is its well-established German short name; or by "*Contributions*", to mirror this use and encourage an easier way to refer to the work in English.

M. Lebech (✉)
Maynooth University, County Kildare, Ireland
e-mail: Mette.lebech@mu.ie

M. Lebech (ed.), *Stein and Husserl's Phenomenology of the Social in Ideas II*, Women in the History of Philosophy and Sciences 25,
https://doi.org/10.1007/978-3-032-17339-3_12

position, we can be sure they testify to her position before she began the work and after she completed it, respectively. The second of those—as it states in the title—*contributes* to the phenomenological discussion of the foundation of psychology and the humanities (issues which were also central to *Ideas II-III*). As such, it can and in fact must be read—in so far as it marks a departure from Husserl's position—as a critique of this latter, both in *Ideas II* and in his unpublished work of the 10 years Stein covered as she put order on his manuscripts and edited some of them (Stein, 2000, p. 2). The continuity between Stein's early work on empathy and her critique of Husserl in *Philosophy of Psychology and the Humanities* allows us to see that during the process of editing *Ideas II*, Stein continued to develop a position which she discovered Husserl did not share regarding a matter central to *Ideas II*—namely, that concerning the constitution of the psyche. Understanding this position allows us to hypothetically identify the conflicting voices of the 1952 edition and thereby open a path for further research into *Ideas II* and into the subject matter treated by that work.

We shall first discuss *On the Problem of Empathy* with its distinction between causality, psychic causality and motivation as a presupposition for the structure of *Ideas II*. Then we shall discuss Stein's revision of *On the Problem of Empathy* in her *Handexemplar*, as it bears the mark of her intense engagement with Husserl's manuscripts of 1916–18. Finally, we shall discuss *Contributions to the Philosophical Foundation of Psychology and the Humanities* as a critique of *Ideas II* and of Husserlian phenomenology around 1920.[2]

As the object of these concluding observations is to establish what motivates the differing viewpoints of Stein and Husserl to make them more visible in the 1952 edition, we shall not discuss whether Stein's "internal" critique is warranted. In so far as this question forms part of the legacy of the work, we shall leave that for future research in phenomenology to discuss.

12.2 On the Problem of Empathy (1917) as a Presupposition for Ideas II

Stein's impression as a student around 1913–14 was that empathy had not really been thematised by Husserl (Stein, 1986, p. 269).[3] She recalls perceiving there to be a gap in its regard, which she could fill with her doctoral dissertation. When she

[2] These considerations were first discussed during an Erasmus+ module on *Ideas II* and *On the Problem of Empathy* co-taught with Henning Peucker at the University of Paderborn in June 2025. I am very grateful to him for sharing his expertise on Husserl, and to the students for their critical engagement.

[3] "In his course on nature and spirit, Husserl had said that an objective outer world could only be experienced intersubjectively, i.e., through a plurality of perceiving individuals who relate in mutual exchange of information. Accordingly, an experience of other individuals is a prerequisite. To the experience, an application of the work of Theodor Lipps, Husserl gave the name *Einfühlung* [Empathy]. What it consists of, however, he nowhere detailed. Here was a lacuna to be filled;

wrote about this in her memoirs in the early thirties (Stein, 1986, 2002), she still indicated that there was such a gap. That she did not revise or relativise that point of view indicates she still thought this gap did obtain in those early years, irrespective of whether she might also have thought or indeed did think that Husserl's later *Cartesian Meditations* would have contributed towards filling it (Stein, 2014).

The extant chapters of Stein's doctoral thesis (Stein, 1917, 1989) testify to her distinction between psychic causality and motivation, in that its third and fourth chapters respectively discuss the constitution of the psychophysical individual and the understanding of the spiritual (i.e., motivated) person. *Ideas II*'s structure (Husserl, 1952, 1989)—while building on Husserl's distinction between nature and spirit—also has a characteristic middle part on the constitution of animal nature. This section is the one in which Stein's editorial activity is most strongly attested. According to Marly Biemel (Husserl, 1952, p. xvii), Stein "put the text beginning on p. 143 together from scratch. From this point onwards there were for the second part no continuous manuscript [by Husserl] available."[4] Stein's intensive editorial activity thus relates to "The Constitution of Psychic Reality Through the Body" (Chapter 3) and "The Constitution of Psychic Reality in Empathy" (Chapter 4) of the Second Part: "The Constitution of Animal Nature." Judging from the concentration of Stein's engagement on these chapters, it looks as if Stein worked hardest where her own contribution was strongest, namely on the constitution of the psyche. She had proposed in her doctoral thesis that the act of empathy contributes indispensably to the constitution of the psyche, despite the psyche forming part of nature in the sense of it being transcendent and being ruled by (psychic) causality, while at the same time supporting the channelling of motivation into feeling and expression. Stein had thus carved out for the psyche a specific, own domain, which is neither that of nature nor that of spirit, but lies in between the two. In so far as Husserl tended to identify psychic causality with motivation,[5] the independence of the object of psychology did not seem clear to him—he alternately classifies psychology under the natural sciences (Husserl, 1973, p. 195) or as in essence transcendental philosophy (Husserl, 1963/1976, *Krisis*, III, B). That the psyche nevertheless *grosso modo* exists in the space between nature and spirit, perhaps overlapping either or both, would have been—again, *grosso modo*—acceptable to him (Husserl, 2025). This is how Stein's clear distinction between causality, psychic causality and motivation could reinforce (or

therefore, I wished to examine what empathy might be. The master found this suggestion not bad at all."

[4] Stein „stellte den Text von S. 143 ab ganz neu zusammen. Von hier ab lag für den zweiten Abschnitt kein zusammenhängendes Manuskript mehr vor." My translation.

[5] I have this idea from Henning Peucker, whose instinctive certainty in the matter I take to indicate he has used the hypothetical identification successfully for many years in his interpretation of Husserl. Bernhard Rang talks about the "substitution of the concept of causality by the concept of motivation" (Rang, 1973, II, 3). I have not, however, been able to independently verify it nor find textual confirmation of it. It does seem to be the case, though, that there exists no systematic distinction between psychic causality and motivation in in Husserl; something to which, of course, Stein's critique also attests.

maybe indeed did *en*force) the systematic structure of *Ideas II*, before Husserl had finished making up his mind about the psyche's exact nature.

In the years following 1910/11, we find in Husserl a division of the sciences, which branches out from the natural sciences because "we all originally tend towards naturalism. We have all a first, original and natural sphere of knowledge: nature in the widest sense of space/time determined real being, the world. To it relates the natural sciences [later changed to 'world-sciences'],[6] the sciences of physical and psychic nature (the sciences of nature in the narrow sense and of the spirit (of spirits, spiritual communities, spiritual production etc.))"[7] (Husserl, 1973, p. 195).

To Stein (1917), who makes less room for such "natural" tendency, we see not only a clear distinction between psyche and spirit (Stein, 1917, pp. 101–2; 1989, p. 91), but also one between psyche and nature, in so far as the psyche is constituted through empathy as alive, while nature as such is not (Stein, 1917, III). In consequence, she understands the division of the sciences to be tripartite, such that the natural sciences, psychology and the humanities respectively study nature (that which is ruled by natural causality), the psyche (that which is characterised by psychic causality) and the spirit (which is characterised by the lawfulness of motivation).

Since both Husserl's and Stein's understandings of the division of the sciences had to be accommodated in the structure of *Ideas II*, it must have come about as a (probably tacit) compromise between them, not least because both authors very likely thought they were in agreement. Although the discrepancy between them may look initially inconsequential, it reveals itself on closer inspection to be of considerable significance for the precise constitution of the psyche and animal nature. At the root of the tension lie their differing evaluations of the importance of empathy for the constitution of the body, and the consequent importance of the originality of the psyche. To Stein, without empathy I would not—I *could* not—see or read the body as the "body of someone" (Stein, 1917, 1989, III), and I therefore could not identify psyche without empathy. To Husserl, in contrast, I seem to identify psyche *because* I identify body; and in relation to the body, I apperceive the soul, thus "appresenting" the other. (Husserl, 1963/1976, *Cartesianische Meditationen*, § 50). They clearly arrive from opposite directions, but might they still arrive in the same place?

In Husserl's later 1927 lectures on nature and spirit (Husserl, 2001), the dichotomy between nature and spirit is not softened by a *tertium quid*. This is fairly strange, given that the same lecture series in 1919 finishes on a lengthy discussion of "The Problems

---

[6] The editor Michael Weiler's remark is significant, since "world-sciences" would then include the natural sciences relating to the natural world and the humanities relating to the life-world: a schema that is echoed in Husserl's later work in competition with another dualism: the tension between naturalistic objectivism and transcendental subjectivism (Husserl, 1963/1976, *Krisis*). It seems significant that Husserl at this point is proposing to classify psychology as a natural science. Stein will not follow him in this.

[7] „Wir haben alle eine erste, ursprüngliche, natürliche Erkenntnissphäre: Die Natur im weitesten Sinn des raumzeitlich bestimmten realen Daseins, die Welt. Ihr entsprechen der Naturwissenschaften [Herausgebers Anm.: 'später zur "Weltwissenschaften" geändert'], die Wissenschaften von der physischen und psychischen Natur (die Wissenschaften von der Natur im engeren Sinn und von dem Geiste (von den Geistern, von den geistigen Gemeinschaften, geistigen Erzeugnissen etc.)." My translation.

Regarding the Psychophysical and Psychical Sphere as well as the Corresponding Sciences" (Husserl, 2002, pp. 211–20). Thus, it would appear that around the time of Stein's redaction of *Ideas II*, Husserl had "more time" for what Stein attempted to do regarding the constitution of the psyche between nature and spirit than he seems to have had later. Michael Weiler, the editor of the two lecture series, explains it by Husserl being henceforth more interested in the theory of science than in the constitution of the regional objects.[8] He concludes: "In the here published [1919] lecture series Husserl proposes [...] comprehensive reflections regarding a transcendental theory of nature and spirit, in which he predominantly makes physical nature and its adequate sciences the object of his phenomenological investigations." (Husserl, 2002, pp. viii–ix). Maybe Husserl's initial concentration on the natural sciences reflects an overall understanding that knowledge "starts with the senses," which he may in fact largely share with Stein. And maybe Husserl's later seeming rejection of the psyche constituting its own region stems from his related insistence that the body "comes first", such that the psyche is "visible" directly in it. This would match with his not regarding empathy as playing any role in the psyche's constitution that it does not already play in relation to the intersubjective constitution of nature in general, and with his reduction of empathy to apperception (or appresentation) in *Cartesian Meditations*. Rang presents Husserl's development as continuously engaged with the opposition of causality and motivation and a search for the "motivational foundation of the concept of causality" (Rang, 1973, p. 5, p. 2, pp. 233–44). According to this account, Husserl would have reverted back to a default position after his encounter with the Steinian critique, because it was the expression of a deeper search for him.

It remains true to say that the second part (The Constitution of Animal Nature) of Biemel's edition of *Ideas II* is more clearly marked by Stein's activity as an editor. That Stein is also more active in asserting this second part's necessity seems to be indicated by the fact that Husserl would subsequently tone down the psyche's characteristic independence, continuing to assimilate it with either the spirit or the body. In *Ideas II*, we find Husserl's manuscripts to form the basis of these chapters, but we also find them taken account of in an understanding of empathy that is likewise found in Stein's *On the Problem of Empathy,* which takes empathy to be an act *in and through which* the psychophysical individual and the person are constituted *as*

[8] „Darüber stellt sich das Jahr 1919 aus systematischer Perspektive einen wichtigen Markierungspunkt für das ‚Natur und Geist'-Thema insofern dar, als Husserl sich in seinen Texten zu ‚Natur und Geist' von nun an mehr und mehr von den zuvor in Zentrum seines Interesses stehenden Problemen der Konstitution abwendete und zunehmend wissenschaftstheoretische bzw. Wissenschaftskritische Fragestellungen, zum Teil auch schon in Verbindung mit kulturkritischen Aspekten, in den Vordergrund der Untersuchungen rückte. So besteht die zentrale Bedeutung der Vorlesung von 1919 gerade in ihrer systematische Übergangsstellung zwischen den vorherigen, mehr an Analysen zur Konstitution von Natur und Geist ausgerichteten Texten aus dem Umkreis der *Ideen II* und den späteren, die ‚Natur und Geist'-Thematik eher wissenschaftstheoretisch angehenden Arbeiten wie in extremster Form der Vorlesung von 1927. In der hier veröffentlichten Vorlesung stellt Husserl [...] umfassende Überlegungen zu einer transzendentalen Theorie von Natur und Geist an, in denen er vornehmlich die physische Natur und die ihr adäquate Wissenschaften zum Gegenstand seiner phänomenologischen Untersuchungen macht."

internally alive and experiencing subjects with the distinct dimensions of psyche and spirit.[9]

It seems plausible that it is only as Stein worked her way through the manuscripts that she identified what she thought was missing in Husserl's account, and thus also therefore what her critique of this account would be. It also stands to reason that their subsequent respective developments would reflect their respective earlier theoretical commitments. That Husserl may not have used the term "psychic causality" completely consistently or have had a clear concept of it would thus have prompted Stein's working out of the distinction between it and motivation in *Beiträge*, even as she had already made this distinction in *On the Problem of Empathy* without insisting on it or pointing it out as a departure from Husserlian intuitions.[10] In *On the Problem of Empathy*, Stein already made room for the experience of sentient or psychic contagion—a phenomenon Husserl does not seem to have grappled with at any point, then or later. In so far as he did not feel obliged to take account of this phenomenon, which so clearly foregrounds the mechanisms of the psyche and highlights it as distinct from motivation, this may help to account for his lack of conviction that the psyche would have to be understood as being distinct from *both* the body and the sphere of motivation.

12.3 Stein's Revisions of on the Problem of Empathy in her Handexemplar

It is probably after Stein had worked on Husserl's manuscripts that she undertook a revision of her part published doctoral thesis, possibly with a view to having a revised edition published or to simply clarify to herself what she had learnt from Husserl regarding empathy during her *Assistenztätigkeit*. The revision, the so-called "*Handexemplar*", is now published as *Zum Problem der Einfühlung* in ESGA 5 (Stein, 2008), and for this reason is often, unfortunately, confounded with her 1917 published thesis. Since Stein would hardly have had time to do the revision while working for Husserl as his assistant, she probably finalised it after her assistantship had ended, and possibly in preparation for (or in tandem with) writing her critique, the *Beiträge*. It seems that the *Handexemplar* brings *On the Problem of Empathy* into closer alignment with Husserlian conceptuality, yet without altering the substance of Stein's understanding.

Although the revision displays a greater closeness to Husserlian phraseology, it also more clearly distances itself from it. "Apperception", for example, is introduced

---

9 We do not quite find this understanding in *Cartesian Meditations*.

10 Since she thought the distinction to be essential (i.e., relying on differences in *eidos*), she would have expected him to have insight into it. It would not have occurred to her before the collaboration faltered that he would not be sufficiently interested in that insight to follow it through to its consequences.

to clarify its relationship with (and distinguish it from) empathy, in order to clarify the phenomenon of psychic contagion and its relationship with empathy.

In the originally published version of *On the Problem of Empathy*, Stein writes:

> At most we could have concluded the presence of the foreign experience from a feeling in ourselves which required the foreign experience to explain its lack of motivation. But thus, we would only have gotten the knowledge of, not a "givenness" of, the foreign experience, as in empathy.[11]

In the *Handexemplar*, this passage has been replaced with the following:

> It is however imaginable that we through the transfer of such a feeling, which is not motivated in our own living, gain experience of its existence in another, that we experience it to be "caused from the outside". This experience would relate to the case of empathy with the foreign other as, in the area of outer experience, causal apperception relates to perception; (like one perceives a hammer stroke on the basis of the impression it has left behind). It would sort under the widest sense of empathy as experience of foreign experience but not be seen as the normal and very common occurrence. (My translation).[12]

In Stein's *Handexemplar* (published for the first time in 2008), psychic contagion and emotional transfer thus form part of empathy (as experience of foreign consciousness) in the very widest sense. One cannot say that about *On the Problem of Empathy*. In the originally-published thesis, these fall outside of empathy, while being adjacent to it. The change in the *Handexemplar* seems prompted by wanting to think about whether apperception could contribute to clarify what empathy is; a desire that could be motivated by Husserl's use of the idea to "replace" empathy through "appresentation", which will figure so prominently in *Cartesian Meditations*.

Stein would not abide by this "inclusion" of apperception and sentient contagion under empathy, however. That she changes her mind may be to put it too strongly – the use of the term apperception for or in relation to empathy just fades into the background. In *Contributions*, we find an entire excursus on psychic contagion (Stein, 2000, 2010, II, I, §3, c), where its specificity is clarified in the context of its social consequences. Even if she here understands there to be in psychic contagion a "sensing" ("*Spüren*") of foreign consciousness, she does not treat it as a kind of empathy.[13]

---

[11] „Höchstens könnten wir sein Vorhandensein aus dem Vorhandenseins des Gefühls in uns, für das wir seiner Motivlosigkeit wegen eine Erklärung brauchen, erschließen. Damit würden wir aber nur ein Wissen, keine ‚Gegebenheit'des fremden Erlebnisses gewinnen wie in der Einfühlung."(Stein, 1917), II, § 5, b, p. 25.

[12] „Es ist allerdings denkbar, dass wir vermittels der Übertragung eines solchen Gefühls, das in unserm Erleben nicht motiviert ist, Erfahrung von seinem Vorhandensein in andern hätten, dass wir es als ‚von außen bewirkt' erleben. Diese Erfahrung verhielte sich aber zum Fall der Einfühlung in das fremde Äußere wie im Gebiet der äußeren Erfahrung die kausale Apperzeption (etwa die Auffassung eines Hammerschlages auf Grund des Eindrucks, den er hinterlassen hat) zur Wahrnehmung. Sie würde sich dem weitesten Bereich der Einfühlung als Erfahrung von fremdem Bewusstseinsleben einfügen, ist aber durchaus nicht als der normale und überall stattfindende Hergang anzusehen."(Stein, 2008, II, § 5, b, p. 37).

[13] Psychic contagion and emotional transfer must be distinguished from empathy in so far as they are available only by means of reflection. In empathy, foreign experience is given "live".

12.4 Contributions to the Philosophical Foundation of Psychology and the Humanities as a Critique of *Ideas II* and of Husserlian Phenomenology around 1920

In the concluding remarks of *Contributions*, entitled "The distinction in principle between psychic and mental being, psychology and the humanities",[14] Stein does precisely what the title announces: distinguish in principle between psyche and spirit, psychic causality and motivation, while also distinguishing both from consciousness (Stein, 2000, 2010, pp. 295–314):

> What counts as a "psyche" is an essential substance, complete in itself with a process of occurrence, which is determined by another kind of essential being and occurring (i.e. physical nature). While the psyche rests upon physical nature, it is only built upon that network, not included in it. [...] All *purely* sensate occurrence is restricted to the life of an isolated individual. There may well be an influencing of sensate occurrence within an individual by other sensate occurrences; but there is no super-individual [psychic] network. *Shared* sensate reality is possible only in so far as the sensate occurrence is a realisation of *mental life*. *Mind* is a going out from yourself, an openness, in a twofold sense: openness for an objective world, which is *experienced*; and openness for someone else's subjectivity, someone else's mind, along with which the objective world is experienced and *lived* in common (Stein, 2000, p. 295).[15]

Thus, although the psyche is essentially built upon nature (through the body), it is distinct from it, and therefore not reducible to it; the form of causality that it is marked by is experienced or experiential causality. If I want to "measure" it, I must "measure" it *in* subjective experience. On the other hand, the psyche is also essentially distinct from spirit, since I cannot share psychic reality with others except by means of motivated mental activity.

Earlier, Stein had characterised motivation in the following manner in the section named "Motivation and the Basic Lawfulness of Mental Living" (I, III, § 1):

> The pointing-yourself-at that we're now remaking, the *intentio* that arises on the basis of immanent data, indicates a new class of experiences, of units that constitute themselves in the current: the class of "apprehensions" or "acts". *Mental living* begins with them. [...] In

[14] The translators of *Philosophy of Psychology and the Humanities*, Catherine Baseheart and Marianne Sawicki, translate "*psychisches*" with "sentient" and "*geistiges*" with "mental". As these are commonly referred to alternative translations to those I use—"psychic" and "spiritual"—I have not changed them.

[15] „Als ‚Psyche 'gilt uns ein in sich geschlossener Seinsbestand und Geschehensverlauf, der evtl. Bedingt ist durch andersartiger Sein und Geschehen (die physische Natur), darauf ruht, aber nur auf diesen Zusammenhang aufgebaut und nicht in ihr einbezogen ist. [...] Alles *rein* psychische Geschehen ist auf das Leben eines isolierten Individuums beschränkt. Wohl gibt es eine *Beeinflüssung* des psychischen Geschehens innerhalb eines Individuums durch anderes psychisches Geschehen, aber keinen überindividuellen Zusammenhang. *Übergreifende* psychische Realität ist nur möglich, soweit das psychische Geschehen Realisation *geistigen Lebens* ist. *Geist* ist Herausgehen aus sich selbst, Offenheit in einem doppelten Sinne: für eine Objektwelt, die *er*lebt wird, und für fremde Subjektivität, fremden Geist, mit dem gemeinsam *er*lebt und *ge*lebt wird."(Stein, 2010, pp. 246–7).

> the realm of acts, we confront new means of connection that we haven't yet encountered up to now. If the gaze points itself successively at a series of continually subsiding data, or rather points on through the data at "external" objectivities, then we have not only a succession of detached apprehensions of a single shape, but *one continuous apprehension*, an *appending* of the later to the earlier ("*apperception*"), a *combination* of single apprehensions ("*synthesis*") and a being-set-in-motion of the later by the earlier ("*motivation*"). [...]. If we designate the connection of acts that we have in view here quite broadly as *motivation*, then we're aware of departing from the customary linguistic usage which restricts this expression to the area of "free acts", especially of willing. [...] Motivation in our general sense, is *the* connection that acts get into with one another: not a mere blending like that of simultaneously or sequentially ebbing phases of experiences, or the associative tying together of experiences, but an *emerging* of the one *out of* the other, a self fulfilling or being fulfilled of the one *on the basis of* the other *for the sake of* the other (Stein, 2000, 39–41).[16]

We see here that Stein's understanding of motivation is expanded to characterise all mental or spiritual activity or life, independently of whether this life is also psychical. Psyche, in contrast, is characterised in the following way (I, II, § 1):

> Up until now, we were speaking about *feelings* of living and *states* of living. Closely examined, the two expressions don't signify the same thing. A conscious being's states of living are wont to assert themselves in the manner peculiar to consciousness. Such a consciousness of life's status, its being experienced, is a feeling of living. [...] Determinations of a reality, its status and properties, manifest themselves as immanent contents in the life feelings – just as in the extra-egoic data. [...] Of course, the ego that is in possession of this real property [life power] shouldn't be confused with the pure ego originally experienced as point of radiation of pure experiences. The ego is grasped only as a bearer of its properties, as a transcendent reality that comes to givenness by manifestation in immanent data but never becomes immanent itself. We shall designate this real ego, its properties and statuses, as the *sentient*. We now see that consciousness and sentience are distinguished from one another in their basic essences: consciousness as realm of "conscious" pure experiencing, and sentience

[16] „Das Sich-richten-auf, dass wir nun meinen, die Intention, die sich auf dem Grunde der immanenten Daten erhebt, bezeichnet eine neue Klasse von Erlebnissen, von Einheiten, die sich im Strom konstituieren: die Klasse der 'Auffassungen' oder 'Akte'. Mit ihnen beginnt das *geistige Leben*. [...] Im Bereich der Akte treten nun neue Verbindungsweisen auf, denen wir bisher noch nicht begegnet sind. Richtet sich der Blick nacheinander auf eine Reihe kontinuierlich ablaufender Daten oder vielmehr durch sie hindurch auf ‚äußere'Gegenständlichkeiten, so haben wir nicht nur ein Nacheinander getrennter Auffassungen einzelner Bilder, sondern *eine durchgehende Auffassung*, ein *Hinnehmen* des Späteren zum Früheren (‚Apperzeption'), ein *Zusammenfassen* der Einzelerfassungen (‚Synthesis') und ein *In-Bewegung-gesetz-werden* der späteren durch die frühere (‚Motivation'). [...] Wenn wir die Verbindung von Akten, die wir hier im Auge haben, ganz allgemein als *Motivation* bezeichnen, so sind wir uns bewußt, von dem üblichen Sprachgebrauch, der diesen Ausdruck auf das Gebiet der ‚freien Akte', insbesondere des Willens, beschränkt. [...] Motivation in unserm allgemeinen Sinn ist *die* Verbindung, die Akte überhaupt miteinander eingehen, kein bloßes Verschmelzen wie das der gleichzeitig oder nacheinander abfließenden Phasen des Erlebnisflusses oder die assoziative Verknüpfung von Erlebnissen, sondern ein *Hervorgehen* des einen aus dem andern, um des andern willen. Die Struktur der Erlebnisse, die allein in das Verhältnis der Motivation eintreten können, ist für das Wesen dieses Verhältnisses durchaus maßgebend: daß Akte ihren Ursprung haben im reinen Ich, von ihm phänomenal ausgehen und hinzielen auf ein gegenständliches. Der ‚Drehpunkt'gewissermaßen, an dem die Motivation ansetzt, ist immer das Ich. Es vollzieht den einen Akt, *weil* es den andern vollzogen hat." ESGA 6, 35–36.

as a sector of transcendent reality manifesting itself in experiences and experiential contents (Stein, 2000, pp. 21–4).[17]

If we interpret all of this as a critique of Husserl's position in 1920, we get that Stein affirms that:

1. The psyche cannot at the same time be consciousness and transcendent. Therefore, it must be distinguished from consciousness.
2. Motivation cannot at the same time be pure I activity and be marked by (psychic) causality. Therefore, it must be distinguished from psychic causality.
3. The psyche is ruled by its own lawfulness: psychic (experienced) causality: It is constituted from all acts in which life power plays a role, but it is distinct from those acts; and their subject, the psychic I, is distinct from consciousness as such, which is imaginable as purely personal.

12.5 Conclusion

To Stein, the distinction between psyche and spirit is essential in order to adequately describe psychic and social life. Her argument for it involves showing how confounding the two does not allow us to have consistent concepts of either. Having had agreement from Husserl to build *Ideas II* on the distinctions and then been unable to carry the argument through must have been frustrating and explains why Stein took her leave. That Husserl was not convinced in the first place explains why Husserl could not publish the result, now bearing the mark of Stein's frustration in its very structure. Whether Husserl grasped her criticism, and what he thought of it if he did, has not to my knowledge so far been explored by anyone. Such exploration would form an adequate backdrop against which one could ask whether Stein's criticism was, or is, justified and/or relevant.

[17] „Wir sprachen bisher von Lebensgefühlen und Lebenszuständen. Genau besehen bedeuten beide Ausdrücke nicht dasselbe. Die Lebenszustände bewußter Wesen pflegen sich bewußtseinsmässig geltend zu machen, und ein solces Bewußtsein von einer Lebenszuständlichkeit, ihr Erlebtwerden, ist ein Lebensgefühl. […] In den Lebensgefühlen als immanenten Gehalten bekunden sich – ähnlich wie in den ichfremden Daten – Beschaffenheiten einer Realität, ihre Zuständlichkeiten und Eigenschaften. […] Das Ich, das im Besitz dieser realen Eigenschaft [*Lebenskraft*] ist, darf natürlich nicht verwechselt werden mit dem reinen Ich. Es ist nur als Träger seiner Eigenschaften erfaßt, als eine tranzendente Realität, die durch Bekundung in immanenten Daten zur Gegebenheit kommt, aber niemals selber immanent wird. Wir werden dies reale Ich, seine Eigenschaften und Zuständlichkeiten als das Psychische bezeichnen und sehen nun, daß Bewußtsein und Psychisches grundwesentlich voneinander unterschieden sind: Bewußtsein als Reich des ‚bewußten' reinen Erlebens und das Psychische als ein Bereich der sich in Erlebnissen und Erlebnisgehalten bekundenden tranzendenten Realität."(Stein, 2010, pp. 21–2).

References

Husserl, E. (1952). *Ideen zu einer reinen Phänomenologie und phänomenologischen Philosophie, Zweites Buch: Phänomenologische Untersuchungen zur Konstitution, Hua IV* (M. Biemel, Ed.). Martinus Nijhoff.

Husserl, E. (1963/1976). Cartesianische Meditationen. *Krisis*, Gesammelte Schriften 8. Felix Meiner.

Husserl, E. (1973). *Zur Phänomenologie der Intersubjektivität: Texte aus dem Nachlass. Erster Teil: 1905–1920, Hua XIII* (I. Kern, Ed.). Martinus Nijhoff.

Husserl, E. (1989). *Ideas pertaining to a pure phenomenology and to a phenomenological philosophy. Second book* (R. Rojcewicz & A. Schuwer, Trans.). Kluwer.

Husserl, E. (2001). *Natur und Geist: Vorlesungen Sommersemester 1927, Hua XXXII* (M. Weiler, Ed.). Kluwer.

Husserl, E. (2002). *Natur und Geist. Vorlesungen Sommersemester 1919, Hua Materialienbände IV* (M. Weiler, Ed.). Kluwer.

Husserl, E. (2025). *Ideen zu einer reinen Phänomenologie und phänomenologischen Philosophie. Zweites Buch: Phänomenologische Untersuchungen zur Konstitution sowie das Nachwort zu meinen Ideen. Die drei Teilentwürfe und Thematische Ausarbeitungen (1908-1930). Hua IV–V* (D. Fonfara, Ed.). Springer.

Rang, B. (1973). *Kausalität und Motivation. Untersuchungen zum Verhältnis von Perspektivität und Objektivität in der Phänomenologie Edmund Husserls.* Martinus Nijhoff.

Stein, E. (1917). *Zum Problem der Einfühlung*. Buchdruckerei des Waisenhauses.

Stein, E. (1986). *Life in a Jewish family, CWES 1.* ICS Publications.

Stein, E. (1989). *On the problem of empathy, CWES 3.* ICS Publications.

Stein, E. (2000). *Philosophy of psychology and the humanities, CWES 7.* ICS Publications.

Stein, E. (2002). *Aus dem Leben einer jüdischen Familie, ESGA 1.* Herder.

Stein, E. (2008). *Zum Problem der Einfühlung, ESGA 5.* Herder.

Stein, E. (2010). *Beiträge zur philosophischen Begründung der Psychologie und der Geisteswissenschaften, ESGA 6.* Herder.

Stein, E. (2014). „'Husserl's tranzendentale Phänomenologie' (1931). Besprechung von: Edmund Husserl, Méditations Cartésiennes und Eugen Fink, Beiträge zu einer phänomenologischen Analyse der psychischen Phänomene" and „Besprechung von: Edmund Husserl, *Die Krisis der europäischen Wissenschaften und die transzendentale Phänomenologie* (1937)" In *'Freiheit und Gnade' und andere Beiträge zu Phänomenologie und Ontologie (1917–1937), ESGA 9*, (pp. 159–161 and 189–193). Herder.

Mette Lebech is Associate Professor of Philosophy at Maynooth University, Ireland, since 1998. She holds degrees from the universities of Copenhagen, Louvain-la-Neuve and Leuven. She has published widely on human dignity and the philosophy of Edith Stein. Her publications include *On the Problem of Human Dignity. A Hermeneutical and Phenomenological Investigation* (Königshausen und Neumann, 2009); *The Philosophy of Edith Stein. From Phenomenology to Metaphysics* (Peter Lang, 2015); *European Sources of Human Dignity. A Commented Anthology of Texts on Human Dignity* (Peter Lang, 2019) and *Value Phenomenology. Taking Account of Edith Stein's Contributions* (Bloomsbury, 2026). She is founding President of the International Association for the Study of the Philosophy of Edith Stein (IASPES) and Series Editor of *Edith Stein Studies*, Bloomsbury. She is a Fellow at the Centre for the History of Women Philosophers and Scientists at the University of Paderborn.

Author Index

M. Lebech (ed.), *Stein and Husserl's Phenomenology of the Social in Ideas II*, Women in the History of Philosophy and Sciences 25, https://doi.org/10.1007/978-3-032-17339-3

Subject Index

M. Lebech (ed.), *Stein and Husserl's Phenomenology of the Social in Ideas II*, Women in the History of Philosophy and Sciences 25, https://doi.org/10.1007/978-3-032-17339-3

The manufacturer's authorised representative in the EU is Springer Nature Customer Service Centre GmbH, Europaplatz 3, 69115 Heidelberg, Germany. If you have any concerns regarding our products, please contact ProductSafety@springernature.com

Printed and bound by CPI Group (UK) Ltd, Croydon, CR0 4YY
21/09/2026
02219989-0005